THE SMARTIFICATION OF EVERYTHING

Critical Perspectives in Sciences, Arts, and Society

From the smart phone to the smart home, smartness has become an almost inescapable reality of everyday life. Supposedly intelligent, interconnected technologies, smart systems have taken on their own forms of life, and it is no longer easy to determine who these systems benefit and what their long-term social and ethical implications may be. In twenty contributions spanning the social sciences, humanities, and the arts, *The Smartification of Everything* offers a deep dive into a variety of studies that critically interrogate smartification processes and systems.

The book is edited by experts in science and technology studies (STS), anthropology, and sociology, and is written by academics and artists working across a diverse range of disciplines, including geography, architecture, and urban studies, among others. The volume moves beyond the digital hype cycle around smart cities, artificial intelligence, or the Internet of Things, which presents smart tech as neat spaces of continuous improvement. Instead, the authors illustrate how smartness is partial, messy, and contested, while also situated in specific sociocultural, historical, spatial, and political realities.

The Smartification of Everything questions the potential and the limitations of smart systems and in doing so furthers our understanding of the complex dynamics today among technology, environment, and power.

MASCHA GUGGANIG is a senior lecturer and co-director of the Center of Life Sciences in Society at the Ludwig Maximilian University of Munich.

KELLY BRONSON holds the Canada Research Chair in Science and Society at University of Ottawa where she is an associate professor of sociology.

VINCENT MIRZA is the director of the Research Centre on the Future of Cities and an associate professor of anthropology at the School of Sociological and Anthropological Studies at the University of Ottawa.

If our world and our futures are technoscientific, then how should we organize this world? And how should we understand these futures? Technoscience and Society seeks to provide new analytical tools to do this, as well as new empirical insights into the changes happening around us. The series encourages shorter, punchier scholarly books providing a crossover forum in which both established researchers and new and emerging scholars can present their investigations into the ever-changing relationship between technoscience and society.

Series editor: Kean Birch, York University
Editorial advisory board: Kelly Bronson, Canada Research Chair in Science and Society (Tier II), University of Ottawa; Alessandro Delfanti, Assistant Professor, University of Toronto; Joan Fujimura Martindale-Bascom, Professor of Sociology, University of Wisconsin–Madison; Jessica Kolopenuk, Assistant Professor, University of Alberta; Linsey McGoey Professor, University of Essex; Ruth Muller, Associate Professor of Science & Technology Policy, Technical University Munich; Fabian Muniesa, Professor, Mines ParisTech; Michelle Murphy, Canada Research Chair in Science and Technology Studies and Environmental Data Justice (Tier I), University of Toronto; Shobita Parthasarathy, Professor of Public Policy, University of Michigan; Jathan Sadowski, Research Fellow, Monash University; David Tyfield, Reader in Environmental Innovation & Sociology, Lancaster University; Malte Ziewitz, Assistant Professor, Cornell University

Also in the series:
An Anthropogenic Table of Elements: Explorations in the Fundamental, edited by Timothy Neale, Thao Phan, and Courtney Addison
Digital (In)justice in the Smart City, edited by Debra Mackinnon, Ryan Burns and Victoria Fast

The Smartification of Everything

Critical Perspectives in Sciences, Arts, and Society

EDITED BY MASCHA GUGGANIG,
KELLY BRONSON, AND VINCENT MIRZA

UNIVERSITY OF TORONTO PRESS
Toronto Buffalo London

© University of Toronto Press 2025
Toronto Buffalo London
utppublishing.com
Printed in Canada

ISBN 978-1-4875-5671-6 (cloth)   ISBN 978-1-4875-5674-7 (EPUB)
ISBN 978-1-4875-5672-3 (paper)   ISBN 978-1-4875-5673-0 (PDF)

Library and Archives Canada Cataloguing in Publication

Title: The smartification of everything : critical perspectives in sciences, arts, and society / edited by Mascha Gugganig, Kelly Bronson, and Vincent Mirza.
Names: Gugganig, Mascha, editor. | Bronson, Kelly, editor. | Mirza, Vincent, editor
Series: Technoscience & society.
Description: Series statement: Technoscience & society | Includes bibliographical references and index.
Identifiers: Canadiana (print) 20250228548 | Canadiana (ebook) 20250228645 | ISBN 9781487556716 (cloth) | ISBN 9781487556723 (paper) | ISBN 9781487556730 (PDF) | ISBN 9781487556747 (EPUB)
Subjects: LCSH: Technology – Social aspects.
Classification: LCC T14.5 .S63 2025 | DDC 303.48/3 – dc23

Cover design: Rafael Chimicatti
Cover image: Karachi Urbal Lab Photo Archives

We wish to acknowledge the land on which the University of Toronto Press operates. This land is the traditional territory of the Wendat, the Anishnaabeg, the Haudenosaunee, the Métis, and the Mississaugas of the Credit First Nation.

This book has been published with the help of funding and support from the Research Centre on the Future of Cities.

University of Toronto Press acknowledges the financial support of the Government of Canada, the Canada Council for the Arts, and the Ontario Arts Council, an agency of the Government of Ontario, for its publishing activities.

 Canada Council for the Arts   Conseil des Arts du Canada

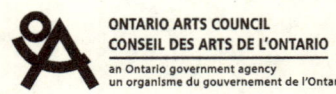 ONTARIO ARTS COUNCIL
CONSEIL DES ARTS DE L'ONTARIO
an Ontario government agency
un organisme du gouvernement de l'Ontario

Funded by the Government of Canada   Financé par le gouvernement du Canada

MIX
Paper | Supporting responsible forestry
FSC   www.fsc.org   FSC® C103567

Contents

List of Illustrations and Tables ix

Foreword: Smartification from Shine to Glare xi
RACHEL DOUGLAS-JONES

Acknowledgments xvii

1 Introduction 3
MASCHA GUGGANIG, KELLY BRONSON, AND VINCENT MIRZA

Part One: The (Settler) Colonial Roots of Smartification

2 Introduction to the (Settler) Colonial Roots of Smartification 31
MASCHA GUGGANIG

3 The Smartification of Global Development 33
KENDRA KINTZI AND HILARY FAXON

4 Refusing the Urban Laboratory 45
ABHISHEK VISWANATHAN AND BOBBIE FAN

5 Making Timber "Smart": Architecture, Technology, and Place in the Pacific Northwest 57
MEG WIESSNER

6 Unformattable: The Meaning of Cleared Land in Copy-Pasteable Smart City Technology Transfer from South Korea to Vietnam 67
JUNE YEOREUM KIM

Part Two: Plastering Frictions and Fissures with Smartness

7 Introduction to Plastering Frictions and Fissures with Smartness   81
   VINCENT MIRZA

8 How Not to Define the Smart City   83
   JUNNAN MU

9 From Smart Buildings to Smart Users? Energy Transition to the Test of Parasitic Humans   95
   GABRIEL DORTHE AND LAURE DOBIGNY

10 Out of Thin Air: Vertiplaces   107
   SIMON RABYNIUK

11 What Is Optimized Farming? Exploring the Smartness Mandate in Canadian Agriculture   119
   SARAH MARQUIS

Part Three: Smartification as Boundary Work

12 Introduction to Smartification as Boundary Work   133
   MASCHA GUGGANIG

13 Reframing Smart Urbanism   137
   ALI FARD

14 The Smart Oasis: Smartification as Process   151
   SEBASTIAN BORNSCHLEGL

15 Disposable: Infrastructures of Exclusion in Dharamshala's Smart City   155
   HANNAH CARLAN

16 Wal*Smartification: Considering the Superimposition of Dockless Shared Electric Scooters on Fayetteville, Arkansas   169
   DEVIN SHEPHERD AND JULIETTE WALKER

Part Four: Opening Up Smartness

17 Introduction to Opening Up Smartness   191
   KELLY BRONSON

18 Floodsmart: "Equity in Action" or Equity Inaction? 195
    MARTIN ABBOTT

19 Should I Stay, or Should I Go? Questioning the "Smartness" of
    Intelligent Pedestrian Traffic Lights in Vienna 207
    POUYA SEPEHR

20 Outsmarting Urbanism: Could Leveraging the
    "Right to Be Rural" Produce Alternative Futures? 219
    S. ASHLEIGH WEEDEN

21 Rural Expertise and the Sewer 231
    JEAN HARDY

Part Five: Rethinking Smartness

22 Introduction to Rethinking Smartness 243
    KELLY BRONSON

23 Contesting Smartness in an Unequal City 245
    SOHA MACKTOOM AND AQDAS FATIMA

24 Future Movement Future – REJECTED
    BRUNO MORESCHI AND GABRIEL PEREIRA 263

25 Thinking like a City 269
    CAROLA MOUJAN

26 Unfamiliar Convenient 285
    CLAIRE GLANOIS AND VYTAUTAS JANKAUSKAS

27 Conclusion 289
    KELLY BRONSON, VINCENT MIRZA, AND MASCHA GUGGANIG

List of Contributors 293

Index 301

Illustrations and Tables

Illustrations

1.1 "sm*ART*" exhibition at University of Ottawa, with Gabriel Dorthe watching "Ghosts in the smart Home" by Lindley, Gradinar and Coulton   12
5.1 Family-owned sawmill and cross-laminated timber (CLT) manufacturer, Washington   59
5.2 Peavy Hall Forest Science Complex at Oregon State University   60
5.3 A mural in Mill City, Oregon, down the road from an engineered wood plant   61
5.4 Cross-laminated timber facility   61
5.5 View of Pahto or Mount Adams from within the lands of the Yakama Nation   62
5.6 This robot at a mass timber fabrication shop in Portland   63
5.7 Wood for glulam arches, a centrepiece of Portland's new airport roof   64
5.8 Bookstore display in Portland, Oregon   64
6.1 A piece of cleared land in Thủ Đức City to build a smart city   69
6.2 A screenshot of the city-building simulator game "Cities: Skylines"   73
9.1 Posters in office spaces explaining how to set up temperature, open the windows, etc.   101
9.2 An example of nudge: "Turn off the light … This is not Versailles here!"   102
9.3 "The Groom Is on Strike"   102
10.1 Vertiport Automation System (Vas) Airspace Ov-1 Diagram, 2020   108
13.1 Google's Lenoir data centre in North Carolina   142
13.2 Google's Lenoir facilities in context   143
13.3 The life and death of mobile devices   146

15.1 Sensor-based underground dustbins located throughout the Dharamshala Municipal Corporation 161
15.2 A stray dog lies trapped at the bottom of a "smart" dustbin 163
15.3 Dharamshala Municipal Corporation's dumping site sits atop the hill, above the farmland of several Gaddi families 165
16.1–16.17 Excerpts of Shepherd and Walker's zine *Wal*Smartification: Considering the Superimposition of Dockless Shared Electric Scooters on Fayetteville* 171–187
19.1 The pushbutton box, which is called a *Bettlerampel* or begging traffic light in Vienna 209
19.2 The similarity between the acoustic signal system and the pushbutton box 210
21.1 The Houghton City Center building, located in downtown Houghton, Michigan 232
23.1 "Contesting Smartness in an Unequal City" 246
23.2 Karachi's rapid urbanization 248
23.3 Map of Karachi, Pakistan 250
23.4 Karachi's Central Business District 251
23.5 "Smart" building practices in Karachi following visions of the world-class city 253
23.6 Materials on south sides of buildings with contrasting thermal properties 254
23.7 "Unsmart" vernacular cooling practices in Karachi's markets 256
23.8 "Smart" (left) and "unsmart" (right) greening practices across Karachi 257
23.9 Activity under "unsmart" green cover versus disintegrated "smart" plantations 259
24.1 Screenshot 1 of short film *Future Movement Future – REJECTED* 265
24.2 Screenshot 2 of short film *Future Movement Future – REJECTED* 266
24.3 Screenshot 3 of short film *Future Movement Future – REJECTED* 266
25.1 *Luciole* at Lyon City Design Urban Forum, Lyon, 2015 274
25.2 Conceptual diagram 275
25.3 *[RIP]_Montevideo*. View of the installation at Espacio de Arte Contemporáneo 276
25.4 Interacting with the book 277
25.5 Abandoned textile factory in Poblenou, Barcelona, 2020 278
25.6 *Future Forest Diorama* – concept images, 2021 279
25.7 *Future Forest Diorama* – the herbarium, 2021 280

Table

20.1 Framework for Critiquing Smartification in Rural Contexts 226

Foreword: Smartification from Shine to Glare

RACHEL DOUGLAS-JONES

My first encounters with smartification occurred around 2013, when a vast, jet black building began to emerge on the banks of Copenhagen's central seawater canal. I walked or biked past the construction site daily on my way to work at the IT University of Copenhagen, seeing foundations grow on a site where, since 1941, no fewer than seventy-five architectural proposals have been put forward and dropped (Weiss 2018). I began receiving emails and hearing lunchtime gossip about how it was materializing as a "hub" for smart projects in the city: academics around me buzzed with the prospect of collaborations and PhD projects. When the building eventually opened in 2018, it would host architecture firms, digital start-ups, community spaces, research institutions, and public bodies in a self-described "ecosystem" that today includes the smart city research network (Bloxhub, n.d.).

I describe this now decade-old encounter because it might seem that *The Smartification of Everything* comes into our hands in the afterlife of a buzzword, one which the inattentive cycles of tech hype might by now have left long behind as they galloped on to the next big thing. Yet the chapters of this edited volume, deeply researched and creative, demonstrate the conceptual tenacity, political insidiousness, and global hold that smartification has gone on to acquire. As a collection, this volume makes the case both for continued critical scholarship on its varied global lives and the urgency of describing what smartification looks like when it turns into a lived condition.

Smartification shines. In these pages it shines through brochures, websites, billboards, blog posts, apps, keynotes, municipal plans, and expo visions. Smartification's shine is wrapped tightly into its promises: in Dharamshala, India, it promises cleanliness, rendering anything and anyone deemed "dirty" "external to smartness, and thus disposable" (Carlan this volume, 156); in Houghton, Michigan, it promises wastewater efficiency, temporarily rendering local expertise irrelevant (Hardy, this volume). The chapters tackle smartification's shine by refusing the easy call of its rhetoric. Instead, they examine the complex interpretative work that happens when "smartness" is made present. They look concretely at the ends smartness is made to serve: to what does it become shackled? The varied answers ground the volume in foundational issues of public interest: cooler urban temperatures, more sustainable food, cleaner cities, flood-aware cities, wastewater and energy management, construction materials, transportation, and policing. Smartification takes the mundane and the municipal and makes them shine for investment, grants, policy schemes, and the promise of technologically enabled – and dependent – futures.

As shine intensifies, it draws in corporate interests, conglomerates, insurance brokers, police forces, entrepreneurs, asset managers, investment brokers, and global and regional banks. Now, we risk smartification's glare. Standing in Karachi, Pakistan, with Macktoom and Fatima, readers will experience this glare bouncing off absurd glass buildings, designed in the image of a global city, wholly unsuited to Pakistan's summer weather. As their surfaces scorch at sixty-six degrees centigrade, in the city below, thick old stone walls sit at twenty-four degrees. Just as glaring sunlight blocks our ability to see surrounding environments, blinding us even as we blink and look away, the chapters of this volume offer us a way to critically manage the glare produced in research domains where normative agendas meet technological hype. Guiding us, these scholars hold up an analytical hand to block the metaphorical glare, at least long enough to see the technologies, knowledges, histories, and politics obscured wherever this glare is produced.

We come to see that the list of things smartification's glare hides is long: without a comparative lens we might not see how smartification is entangled in histories of land, whether colonial violence (Marquis, Kintzi and Faxon), land grabs (Viswanathan and Fan), dispossession (Wiessner) cleared land (Kim), or histories of racial segregation in urban America that reappear in today's digital flood-risk mapping (Abbott). Without the glare of smart technology as a singular solution, existing technologies, whether fabrics that cast shade or clay pots that keep drinking water cool, return to view. Throughout the volume, authors' analyses return to what is produced as smartness's counterpart: that which is *not* "smart" and must be overcome, moved on from; that which represents "the past" or that which, despite its efficacy, has not the shine to attract investment as a feasible solution to the problems smartification has been rolled in to address. Thus, central to the volume is the question of whose interests smartification is made to serve. If, as Kintzi and Faxon suggest, we consider smartification as an evolution of what Tanya Li identified as "rendering technical" (2007, 123), then we must also ask, with Li, how problems become "intimately linked to the availability of a solution" (2007, 7).

In a consistent and unambiguous critique, chapters tackle the techno-solutionism embedded in smartification. Case by case, they show how the study of smartness demands that academics take lessons in criticality, lest they themselves – as Viswanathan and Fan cogently warn – become complicit. Stakeholders – particularly powerful actors who expect projects to be driven, framed, and understood solely through their rationalities and goals (Weeden, this volume) – are closely examined: smartification does not shine for all. Chapters confront the exclusivity of smartification, whether the clearing of land or the removal of people from it in the name of smart futures (Kim, Carlan, this volume). Extraction ricochets through the accounts, genre crossing from data-driven corporate business models to public-space extractivism as universities frame cities as "urban laboratories ... devalu[ing] the lived experience, knowledge and collective power of communities" (Viswanathan and Fan, this volume, 51), and onwards to the more familiar extractivist logics of rare-earth metals upon which digital infrastructures depend (Fard, this volume). Readers reaching for sites of resistance, then, will welcome the presence of refusal in the volume, the positive history of "NO" of Moreschi and Pereira, put forward as the "smart" response to smartification.

Where smartification takes hold matters, and this volume encourages us as readers to explore not only the spaces where smart makes relations anew, but also the histories that it carries with it. Chapters take us across an exhilarating breadth of empirical sites, evidencing the scope and variety of projects around the world. From Amman in Jordan to Pittsburgh in the USA, Seestadt in Austria to Yangon in Myanmar, we see the adaptability of smartification discourse. In their contributions, Kintzi, Faxon, and Carlan point to the "global" or "transnational" discourses we see weaving through the chapters, transformed, as Carlan argues, by the everyday practices of power dynamics of unequal worlds (this volume). But it is the localization of practices where this volume excels, from scooters littering pavements in Fayetteville to cold office workers in France, or the "smart" solution for a "dumb" problem of crossing streets in Vienna (chapters and artworks by Shepherd and Walker, Dorthe and Dobigny, and Sepehr, respectively). Projects described in these pages are long term and loose, allowing ethnographers to return after a few years and observe changes; others are short term and focused, sometimes evaporating as quickly as they arrived. The chapters are not bound by regional or national imaginaries of what technology should do, but they do share a time: the climate crisis saturates this volume. It provides both impetus for smartification and for critical scholarship on it, for ours is a time of extreme heat, wet bulb temperatures, and competing land claims. *The Smartification of Everything* situates salvific technologies as a wholly inadequate and unjust response to the task at hand, no matter how "smart" their intervention claims to be.

As a handbook to a disparate field, one in need of synthesis, comparison, and specificity, Gugganig, Bronson, and Mirza have brought together scholarship from multiple disciplines. Dealing productively with "hot air" or hype in imaginaries of smartness is not easy (Hockenhull and Cohn 2021). But the marriage of concrete needs with the hyperbolic promises of digital technologies rewards a range of approaches. Chapters draw on interviews, ethnography, digital ethnography, zoom attendance, discourse analysis, policy documents, and visual media to lay out the arts of technocratic work. They do so in the spirit of what Andera Ballestero's describes as "generative" (2019, 35) when studying projects of improvement; they make space for wonder in the technocratic, wonder which "takes over when knowledge and understanding cannot master what they should" (Ballestero 2019, x).

It is necessary to be precise about the entitles, devices, and agendas smartness becomes tied to, but what reading these chapters together does is demonstrate the power that smartness's ambiguity has. In what Mu (this volume) calls the "definitional impasse," creativity flourishes: those involved claim smartification, using it to make their case for both present and future action, even if it means redefining "smart" as a good road surface (Carlan, this volume). Beginning with abstract definitions misses how smartness gets defined by those deploying it (see Latour and Callon 1981), something which the chapters of this volume so deftly show. Instead of then departing from the headache-inducing definitional work on smartness and smartification, authors take it as an artefact of the field itself. As such, the volume works *against* its core terms, interrogating the worlds built to push them forward. Instead of definitions, chapters unfold questions: How does smartness problematize in order to offer itself as a solution? How are smart projects legitimated, and by whom? How (and who) do they seduce? Upon what – processes, infrastructures, metals, cables, finances – does smartification depend? What does smartification redirect the gaze towards, or away from, and how is that redirection achieved? How does smartification become the only solution, the only future? The contributions show how such questions can be posed, and the kind of research that can begin to offer answers.

We follow smartification narratives to the places where they meet and are transformed, whether that is at the side of a road next to grazing

Kenyan goats or newly cleared land in Vietnam. *We follow* promises made in Zoom rooms and air-conditioned expo venues. *We follow* data in planned drone projects, drones speculatively enrolled as sky sensors, to generate data that makes marketplaces of city skies. *We follow* actors through webinars and presentations, pitches and performances. *We follow* bodies – bodies that are too hot, bodies that are too cold, bodies that sit drinking hot beverages and wearing warm clothes in sensor-laden, temperature-controlled, standardized offices. *We follow* devices, from overflowing underground dustbins with sensors sounding alerts to negligent outsourced contractors to sensors that simply run out of battery; GIS models, machine learning programs, smartphone apps, and digital government portals. As Halpern and Mitchell argue, the technology matters, but smartness is "an epistemology" (2023, xi). So, we follow responses. Finally, the scholars in this volume follow *image*. In addition to foregrounding the role of image-making in smartification worlds, the centrality of artwork and image-making in this volume counters the homogeneity of the aesthetics smartification often deploys, showing what "smart" looks like when lived and speculatively reimagined.

More than a decade ago, the anthropologist Annelise Riles (2011) called for a closer examination of the dreams that take hold in technocratic worlds, and this volume prompts us to ask to whom these dreams of smartification belong. Between them, the collected chapters and artistic contributions of *The Smartification of Everything* excavate the dreams, desires, and fears that propel along technological fantasy and its material instantiations. The volume excels at drawing our eyes to what smartification would, in its device-shine, data-shine, and "smart" shine, help us not to see. Thought of as a set of lessons in handling smartification's shine, extending our analytical perspectives beyond the glittering, given, and broken promises of optimization, the challenge *The Smartification of Everything* lays down for its readers is how to conduct scholarship that is neither seduced by the shine nor dazzled by smartification's glare.

References

Ballestero, Andrea. 2019. *A Future History of Water.* Duke University Press. https://doi.org/10.1215/9781478004516.
Bloxhub. n.d. "Smart City Research Network." Accessed May 13, 2023. https://bloxhub.org/smart-city-research/.
Halpern, Orit, and Robert Mitchell. 2023. *The Smartness Mandate.* MIT Press. https://doi.org/10.7551/mitpress/14623.001.0001.
Hockenhull, Michael, and Marisa Leavitt Cohn. 2021. "Hot Air and Corporate Sociotechnical Imaginaries: Performing and Translating Digital Futures in

the Danish Tech Scene." *New Media & Society* 23, no. 2. https://doi.org/10.1177/1461444820929319.

Latour, Bruno, and Michel Callon. 1981. "Unscrewing the Big Leviathan: How Actors Macro-Structure Reality and How Sociologists Help Them to Do So." In *Advances in Social Theory and Methodology: Towards an Integration of Micro- and Macro- Sociologies,* edited by K. Knorr-Cetina and A.V. Cicourel. Routledge & Kegan Paul.

Li, Tania Murray. 2007. *The Will to Improve: Governmentality, Development, and the Practice of Politics.* Duke University Press. https://doi.org/10.1215/9780822389781.

Riles, Annelise. 2011. *Collateral Knowledge: Legal Reasoning in the Global Financial Markets.* University of Chicago Press. https://doi.org/10.7208/chicago/9780226719344.001.0001.

Weiss, Kristoffer Lindhardt, ed. 2018. *BLOX.* Realdania By & Bygs Forlag.

Acknowledgments

We would like to thank the Social Sciences and Humanities Research Council of Canada, and the University of Ottawa Research Centre for the Future of Cities for supporting the conference and exhibition "The Smartification of Everything" that took place in Ottawa, March 7–11, 2022. At the time, only barely recovering from yet another COVID-19 infection wave (it was Omicron) and lock-down, we were grateful that so many presenters made their way to the snowy capital of Canada. We also want to thank our three keynote speakers, Orit Halpern, Shannon Mattern and Carl DiSalvo, the latter barely making it to Ottawa after he found out his password was about to expire. Many thanks also go to Rachel Douglas-Jones who organised a virtual hub at their ETHOS Lab at IT University of Copenhagen. Special thanks also go to the conference and exhibition organising team, which included our research assistants Nazeeha Jafri and Dana Malapit, which were supported through the University of Ottawa Alex Trebek Forum for Dialogue, as well as Ariane Millette-St-Hilaire, Cristian Maximiliano Cabrera Van Cauwlaert, Adrien Savolle, and Zoe Campbell. Finally, we want to thank all contributors to this edited volume for their endurance seeing this book through.

THE SMARTIFICATION OF EVERYTHING

1 Introduction

MASCHA GUGGANIG, KELLY BRONSON, AND
VINCENT MIRZA

It seems to be an inescapable reality that everything is being made smart: smart fridges and smart homes, in smart cities that are increasingly surrounded by smart farms, smart forests, and even smart fisheries. Smartness has become a ubiquitous, pervasive reality of everyday life. When bodies, cities, gadgets, buildings, or such mundane things as shower heads[1] are described as "smart," smartness is evidently no longer an exclusively human virtue. In fact, smartness is increasingly associated with a steady digitization and wireless connection of various sectors, spaces, and landscapes. It is a trend that is arguably contrary to humanness: smartness seems to gain most meaning when countered with "inefficient" nature and human virtues (Halpern and Günel 2017; Luque-Ayala et al. 2016). Concurrently, "smart" technologies are envisioned to produce a more resilient human species that is "able to absorb and survive environmental, economic, and security crises" (Halpern et al. 2017, 107). This shift from human qualities to precisely "efficient" analysing and executing systems, which in turn are to foster "smart" subjects and virtues, has profound consequences. Smartness is imagined and normalized by specific actors, whose interests are often disguised by the shiny, slick surfaces of smart gadgets, objects, or buildings. ICTs in ostensibly smart systems reshape power relations between the public and private sector with considerable effects on democratic systems (Benessia and Pereira 2015; Foth et al. 2015; Mattern 2021; Sadowski 2020). It is not always evident who is served by smart systems, nor what social justice and ethical issues may arise in the long run (Araya 2015; Burns et al. 2023; McFarlane and Söderström 2017; Sadowski 2020).

---

1 As seen in a display on smart gadgets at the Museum of Communication Frankfurt, May 2023.

While the marketing of products, places, or buildings as "smart" has exploded in recent years, it is not a recent phenomenon. Already in the 1990s, cultural analyst Andrew Ross diagnosed *The New Smartness* (Ross 1993) as "an advanced form of competition in the sphere of intelligence, where knowledge, more than ever, is a species of power, and technology is its chief field of exercise" (108).[2] Three decades later, Orit Halpern and Robert Mitchell take up this argument when they state that beyond conventional critiques of specific genealogies, promises, and perils of "smartness," it is better understood as an epistemology that is reliant on new practices, subjects,[3] and technologies, including artificial intelligence and machine learning as "the bedrock of smartness" (2023, xi). As "intelligent," interconnected technologies, smart systems have thus taken on their own forms of life (Fischer 2003), not only by pervading people's lives but also by normalizing an epistemology in which only some objects, places, and people count as "smart," while others are rendered "dumb" (Halpern and Mitchell 2023). These widely shared associations with AI-driven, data-, network-, and computer-based systems can be referred to more specifically as technoscientific smartness. While it does not inherently stand in opposition to human and other species' forms of intelligence, it indicates a growing trend where anything "smart" is unquestionably rendered technical (see Li 2007). This book is concerned with this pervasive conflation and slippage, as well as how actors distinguish between "smart" and "technoscientific." Through case studies, concepts, and theoretical approaches, the volume's contributions attend to the ambivalences and discrepancies that are defined by actors' opportunistic and political reasonings, as well as sociocultural circumstances.

In critical studies of smartness, one cannot miss a strong urban focus, and this is not accidental. Smart systems originated in the context of cities (Karvonen et al. 2019; Kitchin 2011, 2014; Luque-Ayala et al. 2016; Mattern 2021), more specifically, in big data modelling of cities. "Smart" is often equated with such virtues as "'good,' 'healthy' and technologically advanced" cities and their citizens, yet it lacks a concrete definition (Vanolo 2014, 886; see also Hollands 2008). It often acts as an "empty

---

2 If one were to take Google's Ngram Viewer as an indicator of the prevalence of concepts over time, one can see an exponential jump in the 1980s, following a long period of stable and low frequency since the 1800s (accessed June 7, 2024), https://books.google.com/ngrams/graph?content=smart&year_start=1800&year_end=2019&corpus=en-2019&smoothing=0.

3 New, and thus far less studied subjects are data workers that generate data sets fed into smart systems, such as the "Smart Brain" for Chinese smart city initiatives (Zhao and Douglas-Jones 2022).

signifier" that aligns with influential actors' agendas, from city planners to private investors (Burns et al. 2023). Once again, digital tools are relied upon at the expense of more inclusive interventions, place-based urban planning, policy, and citizen participation (Aurigi and Odendaal 2020; Datta and Odendaal 2019; Gabrys 2014; Rosol and Blue 2022). Further, the growing prominence of smart cities and systems in the Global South in the last decade (Datta 2015; Watson 2013) continues to perpetuate a long history of uneven global development defined by inequity in colonial and capitalist expansions. It is particularly here that one can find simplistic, developmentalist technology transfer narratives and a logic of constant demoing and prototyping, or demo-logics (Halpern and Günel 2017). These narratives and material manifestations raise the need for critical attention to the cultural, politico-economic, and spatial context of this ostensibly universal model, not least to provincialize debates on the Global North (Datta 2023). Since planetary mining for materials to develop "smart" systems and their related scientific knowledges and technical expertise precedes the history of nation-states (Arboleda 2020), the extension beyond the urban – and the Global North – is a worthwhile exploration (Burns et al. 2023, 21). Likewise, visionaries of smartness, or sociotechnical vanguards (see Hilgartner 2015), foster sociotechnical imaginaries that often manifest in a distinct futuristic visual culture and materiality – architectural renderings, billboards advertising smart cities, etc. – that profoundly shape people's everyday lives in the here and now. Artists and artist-researchers create important multisensorial interventions in these dominant visualizations that are often neglected in academia. Expanding upon the urban, the Global North, and the written academic text thus forms the basis for this edited volume as a contribution to critical studies of smartness.

The Smartification of Everything

This edited volume, *The Smartification of Everything: Critical Perspectives in Sciences, Arts, and Society*, aims to rectify this geographic urban and Global North bias in scholarship on smart systems to show the potential of a transdisciplinary dialogue between the arts and sciences. By taking the bounded concept of "smartness" as open to inquiry, going beyond the obsolete urban-rural division too often reproduced in smartness discourses, and engaging with artists and artist-researchers' provocative critiques, the book is rooted in a science and technology studies (STS) and art, science, and technology studies (ASTS) (Rogers et al. 2021) mode of inquiry. Through contributions across various arts and research disciplines – STS, anthropology, geography, urban studies, architecture, and related fields – the book illustrates how smartness is partial, messy, and contested, while bound to specific sociocultural, historical, spatial, and political realities. In their different foci and approaches, the collection poses the following questions: When conceiving of the urban as broader processes and (capitalist) relations than the city (Brenner and Schmid 2011), how has smartness manifested in geographies and sectors beyond and in connection to the urban? What are its (settler colonial) historical roots? What role do supposedly universal, smart systems play in smoothing out friction? What boundaries around technoscientific smartness do actors draw and contest, and what can be learned, when opening up smartness, about their specific cultural, historical, and geographic contexts? Is smartness defined once and for all as the use of highly integrated digital tools, as in the Internet of Things, or should the notion be reclaimed from this digital-tech rhetoric? And, to what extent may algorithms generate new insights and meaning?

The underlying practices and dynamics behind such neatly bound "smart" places, objects, or subjects often remain opaque. Reflecting on Ross's (1993) and Halpern and Mitchell's (2023) approach to smartness as epistemology, we propose that the concept of smartification delineates the rendering (simplifying or obscuring) of complex social relations and environments as technical or "smart," which are compartmentalized into neat objects, subjects, or spaces of continuous improvement. As "no thing, no space, is safe from smartification" (Sadowski 2020, 1), we approach the pervasive *smartification of everything* by attending to smartification as *processes and practices; actors' performative attributions of (technoscientific) smartness to places and entities in their perceived need to transform them into seamless, bounded, and technically interconnected holes which are absolved of human messiness, partiality, and friction.* Analysing smartification processes thus means capturing the mutual process of making spaces, fields, and sectors smart, and how smart systems in turn reflect sociocultural, political, and economic contexts (see Kumar 2019). In other words, what is smartness when forests, oceans, or agricultural fields become the very sources of legitimacy for said smartness? Just as in other technoscientific fields, the way problems are framed is inseparable from the "smart" solutions that their proponents advocate (Gugganig 2025; Hilgartner and Bosk 1988; Li 2007). Tracing these processes and dynamics of the "optimization of everything" (Jacobs 2022), the book thereby also takes a fashionable term from high-tech circles – the internet/smartification of everything[4] – to extend critical accounts, such as to the smartification

---

4 CISCO Corporation, www.cisco.com/c/dam/en_us/about/business-insights/docs/ioe-value-index-faq.pdf; DHL, www.dhl.com/de-en/home/insights-and-innovation/thought-leadership/trend-reports/smartification.html (both accessed December 19, 2022).

of agriculture (Maschewski and Nosthoff 2022); the urban, the rural, and the daily domains; sites in the Global South. Proponents of technical systems, in urban planning, policy, or private investment, often establish norms for what counts as "smart" while disguising their constitutive power when calibrating technoscientific smartness. We are also inspired by Burns et al.'s (2023) approach to "smart" as an empty signifier which gets "filled" according to the interests of influential players in smart cities.[5] Many contributions in this book move beyond the urban to demonstrate how the empty, flexibly adopted virtue of smartness has pervaded urban, peri-urban, and rural spaces, subjects, and objects. On an analytic level, smartness as empty signifier also has a heuristic value: just as much as it is difficult to define smartness, it is equally difficult, and perhaps futile, to define one way to theorize "smartness." We also refrain from standardizing a single approach as the edited volume encompasses both scholarly and artistic perspectives on smartness, the analytical and artistic diversity of which helps us to comprehend practices and performances of envisioning, establishing, maintaining, and contesting smart systems, spaces, and objects – that is, smartification.

With a focus both on research and the arts, the volume also contributes to a growing body of work in STS that puts the arts in conversation with academic work as an epistemological process.[6] The edited volume offers rich empirical case studies and artworks thereof, as well as analytical approaches to these processes, which were presented in the symposium The Smartification of Everything and the art exhibition *smART* at the University of Ottawa and virtually in March 2022.[7] While researchers offer critical analyses of ways in which smartness takes on meaning, artists go a step further by challenging us with what-if scenarios. They provoke us to think about the sociocultural, ethical, and moral dimensions of anything "smart" seemingly improving humankind in often deeply personal and intimate ways. Along this line, we take inspiration from art exhibitions, such as *Touch Deeper*, that present artists' and designers' "alternative views on the efficiency and intelligence of [smart] devices and examine how we, with an increasing number of digital companions

---

5 Following Laclau (2007), they note that as a signifier, smart does not have an inherent meaning but retains meaning through relationality – in ways that actors use, accept, or resist it – revealing its ontological instability and their political power as derived from its emptiness (Burns et al. 2023, 5–10).
6 See for instance the *Making & Doing* Programme by the Society for the Social Studies of Science (4s), accessed June 7, 2025, https://www.4sonline.org/making_and_doing.php.
7 University of Ottawa and online, March 9–11, 2022, see www.youtube.com/@thesmartificationofeveryth5883 for recordings of the talks.

and surroundings, want to live" (Seiner 2016). For instance, the ethics of allowing self-driving cars is approached through a fictional review panel of academics (Moreschi and Pereira, this volume), while the Internet of Things is considered its own new species (Glanois and Jankauskas, this volume). What unexpected forms of relationships arise between humans and smart forests (Moujan, this volume), and what do these differences and similarities say about human and technoscientific smartness?

The remainder of this introduction offers a brief etymological exploration of smartness, an overview of most relevant work, its urban bias and understudied rural relations, and the value of bringing arts and sciences into dialogue over smartness. We subsequently introduce the volume's contributions in five themes on the smartification of everything and provide an outlook on how smartness may be rethought in contemporary times.

From Intelligence to Smartness

When the meaning of smartness has exceeded the virtues of humans, it is worth considering the term's origin. Etymologically, the verb and adjective "smart" has roots in the Middle English term *smerten*, to cause pain, with early seventeenth century references to smart words being "on the impertinent side of witty" and persons being "active, intelligent, clever." Much later, in the twentieth century, its meaning was expanded to devices "behaving as though guided by intelligence."[8] The attribution to things "as though" is key as it underscores a mere likeness to a person's intelligence, rather than a firm list of stand-alone virtues. It also indicates that as a consequence, in the last decades, smartness has become normalized as having technoscientific attributions. Technoscientific systems have become the fetishized standard against which contemporary "smartness" is measured, devoid of the burdensome human that repels full quantification.[9] For Palanca-Castan and colleagues (2021), the humanization in attributing intelligence to technological processes problematically conflates biological and technological qualities, while the latter, such as in AI systems, more adequately reflects "purposeful behaviour." A dominant algorithmic epistemology indicates a

---

8 www.etymonline.com/word/smart, accessed July 19, 2023.
9 Geoffrey Bowker explains that intelligent citizens nowadays cannot read programs running data sets, as "increasingly scientific models are compared primarily against other models. So, let's take the unnecessary human out of the equation" (Bowker 2013, 169f).

growing "fetishisation of information that ascribes super-natural divination to digital technology" (Miles 2019, 5; see also Chun 2011) rather than (if at all) to humans. Such attributions also illustrate a more fundamental anthropocentric logic underlying smart devices and systems in two ways: first, the original connotation of *human* intelligence (rather than, say, non-human animals, plants, etc.), and second, the subsequent anthropocentric design of such systems; what smart systems should serve are human needs and epistemologies – be it to "save" water, to couple one's smartphone to the smart fridge for efficient grocery shopping, or to manage "ecosystems" by turning the environment into the "Internet of Nature" (Galle et al. 2019; see for critiques Gabrys 2022; Sheikh et al. 2023; Yigitcanlar et al. 2019).

With roots in computer science and associations with bombs, chip cards, and a range of concrete objects and abstract notions, the term "smart" also evokes a "salvific promise to restore economic growth and modern welfare" (Benessia and Pereira 2015, 79). This smartness, as Halpern and Günel (2017) argue, is always specific and directly refers to computationally managed systems that have the potential to learn about themselves. In the smart city, such computational scrutiny is "to render the previously opaque or indeterminate not merely knowable but actionable [and] to permit the 'optimization' of all the flows of matter, energy and information" (Greenfield 2013), in other words, to render it technoscientifically smart. In smart city marketing, smartness is often employed in ahistorical, aspatial, and homogenizing ways (Greenfield 2013), while the streamlined global discourse and abstract notion of "the smart city" is multifaceted and anything but stable (Joss et al. 2019). Similarly, in other domains, like in agriculture, "smart" is "ill defined and semiotically gravid" (Miles 2019, 10), and the historical traces the term has left are just as multifarious as the associations it evokes in contemporary times (see Burns et al. 2023).

Urban Bias and Rural Relations

While the most commonly studied phenomenon of smartness is the city, a growing number of scholars show that supposedly smart spaces have manifested far beyond the urban. More fundamentally, prominent urban scholar Henry Lefebvre ([1973] 1976) long critiqued depictions of the city as an entity supposedly independent from rural context, a standpoint that only recently started to provoke a reconsideration of his work (Elden and Morton 2016). Environmental historians and historians of supposedly "rural" spaces have likewise pointed to their deep integration with cities and the construction of rural spaces, like farms, as cities unto themselves (Cronon 1991). Rural extraction sites, data centres, or cable infrastructures, the latter often on settler colonial land, are key for the establishment and maintenance of smart cities, having extraordinary environmental impacts due to the massive need for electricity and water (Burrell 2020; Fard, this volume; Hogan 2018; Levenda and Mahmoudi 2019). As Halpern and Mitchell recently argued, what is essential to any form of smartness is its material precondition, in the massive quantities of copper that are extracted from mines to set up electrical conductive properties "in essentially every machine on earth" (2023, xiv). Smartification processes – the rendering of social relations and environments as "smart" – thus include and expand the city in two ways: both in terms of cities' reliance on rural sites and people, and in how *any* smart system relies upon such sites. This is also well illustrated in the short film *Rare Earthenware* by the artist collective Unknown Fields Division (2015), which traces and repurposes the rare earth elements needed, and the toxic waste produced, in the manufacturing of smartphones, laptops, or electric car batteries (see also Gugganig and Bronson).

Returning to the city, in the 1970s, designers in the Architecture Machine Group (later MIT's Media Lab) integrated computational systems into buildings in their hopes to face urban degradation and racial segregation in the United States and the Global South (Halpern and Günel 2017; Negroponte 1970; Salter 2022).[10] Three decades later, in the early 2000s, a number of urban development projects, including Masdar City in the United Arab Emirates or Portugal's PlanIT Valley, became canonical "forerunners and exemplars of the kind of urban environment we might inhabit once the cities of Earth have been decisively colonized by networked informatics" (Greenfield 2013). Yet this long-imagined, urban pinnacle may never be reached. According to Halpern and Günel, smart cities are never a finished product, as their development "follows a logic of demoing, constant prototyping, testing and updating," with "infinitely replicable but always preliminary versions" of the smart city across the globe (2017, 2). Given their inherent demo-logic, smart cities don't necessarily "fail" since they nevertheless mobilize resources for future investment and perpetuate powerful visions (Günel 2019). Sadowski and Bendor (2019) similarly argue that the "smart city" is not a goal to be reached but a sociotechnical imaginary (advanced by IT companies) that relies upon and further cultivates narratives of urban crises and technological salvation. The smart city then turns into an imagined

---

10 Others trace the origin of the "smart city" to the IT company IBM, where aspirations were similarly for technology to improve the functioning of cities (Jacobs 2022).

solution to various crises beyond the urban (Datta 2015) in the form of the smart subject (Dorthe and Dobigny, this volume; Mu, this volume; Sadowski 2020), smart electrical grids (Slayton 2013), smart agriculture (Bronson 2022), smart forestry (Gabrys 2020; Halpern and Mitchell 2023), or rural development (Hardy, this volume). Hence, smartness becomes a powerful currency and promissory future (see Rajan 2006) that rarely materializes in finished products, but more so in the form of open-ended dreams and visions. As Kintzi and Faxon illustrate in this volume in one of their case studies in Myanmar, organizers of an agritech hackathon were less concerned with creating profitable digital systems and quite content with "merely" keeping the young generation's future interest in anything digital and agriculture.

Our volume draws upon work that points to an analytical bias towards the city and critiques a common "*urban* slant" (Fraser 2019, 896; emphasis original) that is particularly prominent in the context of agri-food systems and rural development (Bronson and Knezevic 2016; Weeden 2022). On the one hand, "standardizing the environment" is required in smart farming (Bronson 2019, 4) as in smart forestry (Gabrys 2020; Halpern and Mitchell 2023), and it often serves a self-fulfilling prophecy: to accord with commercial imperatives of large-scale farming, expensive machinery is required to generate predictable topographies that in turn make smart farming technologies essential (Bronson 2022; Marquis, this volume) and the rural legible to the state (see Scott 2008). On the other hand, given the inherent bias towards the urban in rural development and policy, the rural often turns into a default plan B: "failed cities, cities in waiting, or resource banks for urban demands" (Weeden, this volume, 220). Consequently, to justify "development" – similar to developmental policies in countries of the Global South (Kintzi and Faxon, this volume) – smartification becomes the remedy that often equates to *urbanization* (Weeden, this volume).

smART: Bringing Arts and Sciences Together, over Smartness

This volume brings scientific research in conversation with the arts by considering them as equal forms of knowledge production (Haraway 2016; Rogers 2022). Social science and artistic research have much in common, be it the emphasis on practice as a site of knowledge creation (Borgdorff et al. 2019, 1; Glanois and Jankauskas, this volume; Moreschi and Pereira, this volume) or the virtue of self-reflection among artists and anthropologists questioning their social positioning (Helmreich and Jones 2018, 99; Wiessner, this volume). Further, artistic approaches extend scholarly analyses by disturbing, confusing, and

Figure 1.1. "sm*ART*" exhibition at University of Ottawa, with Gabriel Dorthe watching "Ghosts in the smart Home" by Lindley, Gradinar and Coulton. Image by Mascha Gugganig.

reconceptualizing conventional norms and ideas (see Jones 2005), by discovering and displaying "things that more ordered [academic] processes miss" (Pinch 2021, xxi). Artists generate intimate, visceral provocations, leading the observant to reconsider on a personal, societal, or political level their own and other humans' and species' state of being. Art, as art-technology historian Caroline A. Jones asserts, "can offer those who experience it an intuitive grasp of worldly entanglements and living relations that take decades to fully parse" (2021, xx), thus defying academia's chronopolitics (Felt 2025), its obsession with on-schedule impact factors, teaching evaluations, and a general audit culture (see Strathern 2000). This approach to art and science as equally relevant forms of human knowledge production has most recently been approached through the field of ASTS (Rogers et al. 2021; see also Borgdorff et al. 2019). To grasp the growing emergence of art-science practices and collaborations analytically, ASTS draws on methods and concepts from STS to study art–science as equal contributions of ideas among artists and scientists. In this edited volume, we share the significance of questioning "convenient borders or definitions of those two knowledge communities" (Rogers et al. 2021, 2) by combining scientific research and artwork contributions, authors that are both researchers and artists, or researchers practicing artistic methods (e.g., Moujan; Wiessner).

Accompanying our symposium *The Smartification of Everything* in March 2022, the exhibition *smART* at the University of Ottawa's Department of Visual Arts intended to provide such overlapping artistic and artistic-research perspectives on smartness (fig. 1.1). Fifteen artists, art collectives, and artist-scholars showcased various ways in which technoscientific systems render humans' social, environmental, and virtual relationships with their lived-in worlds, using the multisensorial media of video, digital renderings, and paintings. Such smartness artwork – or, as we called it, smART – has parallels with surveillance and forensic art in how artists are similarly "availing themselves of highly technical investigative tools" to surveil surveillance (Helmreich and Jones 2018, 107) – to surveil smartness. Artists' and artist-scholars' views on smartness are generative for several reasons: as already indicated, they question what constitutes "smartness," not from a purely intellectual perspective but by provoking visceral, affective experiences of anger, fear, wonder, or joy. The multisensorial experience and embodiment of smartness can engender a deeper level of comprehension and reflection than a written, academic text, such as when viewing Bornschlegl's Visual Vignette (this volume), whose display of the cacophony of an (urban) "smartness" is both topic and method, leaving the viewer confused, distracted, or amused. Hence, while human sensorial capacities tend to be rendered obsolete and effectively replaced by "the ultimate automated [cybernetic] organism" of a smart system/city (Gabrys 2016, 253), through multisensorial techniques, artists also formulate a methodological critique on technoscientific smartness. In the short film *Frames* by Madeline Ashby, Farhad Pakdel, and sava saheli singh (2019), also featured in the *smART* exhibition, the viewer is transported into a smart city whose surveillance system minutely tracks, interprets, and logs a woman's actions, visualized through computer-rendered imagery. Watching the visibly disturbed woman, we are provoked by the audiovisual stimuli to ask to what extent we already allow our daily lives to be tracked, and to what extent we have accepted "smart" to mean "surveilled." Further, artistic approaches can visually trace and critique increasingly complex, globally interconnected networks of material extraction that smart gadgets and systems rely on. For instance, the short film *Rare Earthenware* by the artist collective Unknown Fields Division (2015) traces globally extracted rare earth elements and resources for manufacturing smart gadgets, yet it does not stop there. As a conceptual critique, the short film ends with an artistic intervention of turning radioactive waste into ceramic vases, packaged presumably for display in art galleries – or people's homes? How, they seem to ask, do we deal with (read: ignore) the toxic by-products of technoscientific smartness? Photography is another medium with which artists transpose smartification phenomena. In her photo essay *Making Timber "Smart":*

Architecture, Technology, and Place in the Pacific Northwest, artist and media scholar Megan Wiessner uses photography to critique dominant imagery in environmental design, which tends to fetishize techno-natural hybrids. While she is aware of the risky slippage in visuals – between legitimizing and unsettling (visual) knowledge production – this ambivalence can be understood as a powerful datum played back to the viewer: how do they judge the visual aesthetics of a settler colonial, "climate-smart" forestry? Can data-extractivist smartness be beautiful? Can one "see" (settler) colonialism in these instances, and if so, how?

In an effort not to use artwork as fashionable labels that merely illustrate a topic, in this edited volume artistic contributions are entwined, rather than neatly separated from research work, taking inspiration from Richard Misrach and Kate Orff's *Petrochemical America* (2012), Bruno Latour and Peter Weibel's research-exhibition catalogue *Critical Zones* (2020), or Walter Kehm's curated volume *Accidental Wilderness* (2020). We also follow the approach of scholars and artists to question the "labour division" between text as descriptor and visuals as illustration of text, to art research creation in and of itself (Gugganig and Douglas-Jones 2021; see also Arens and Schlünder 2020; Bornschlegl, this volume; Favero 2017). Artwork presented includes a photo essay (Wiessner), a zine (Shepherd and Walker), a Visual Vignette (Bornschlegl), and two short videos (Moreschi and Pereira; Jankauskas and Glanois).

Contributions to the Edited Volume

Organizing fifteen research and five artistic contributions on smartness is not an easy endeavour: shall we order them along methodological, disciplinary, thematic, or geographic parameters? As we aimed to think across urban-rural dimensions, geographically distinct places, and artistic and research approaches, we deliberately opted not to structure, and thus perpetuate, an approach to smartness through the lens of "rural" vs. "urban," "Global South" vs. "Global North," or arts vs. research. Instead, these dimensions and formats run through the book, which is organized along concepts that the respective case studies speak to. We start off with a (settler) colonial critique of smartness in order to highlight its rootedness in long-lasting histories of material and (geo)political exploitation. Part 2 is organized around frictions of smartness as an anthropological concept; part 3 looks at the STS concept of boundary work to maintain and contest technoscience/smartness; part 4 examines another STS concept, the "black box" of technoscience/smartness; and part 5 rethinks smartness to demonstrate novel, often artistic approaches. These sections are, of course, not completely distinct from each other as there are (fruitful) overlaps to think with.

Part 1, "The (Settler) Colonial Roots of Smartification," confronts the colonial and developmental foundation of smartification processes through case studies from Jordan, Myanmar, the United States, South Korea, and Vietnam. Kendra Kintzi and Hilary Faxon start off by offering a framework for critical geographies of smart development to study how the "circuitry of smartification intersects with older flows of capital and knowledge" (34). Through the case of Myanmar's smart farms and Jordan's smart grid system, they demonstrate that just like "development," the appendix of "smart" has turned into an industry buzzword to attract foreign private investors who are motivated by the belief that technologies will fix emerging markets and uplift the poor. Next, Abhishek Viswanathan and Bobbie Fan reveal how Pittsburgh's (US) powerful nexus of universities, foundations, corporations, and politicians advance a "Creative City" and smart/surveillance agenda. This agenda excludes working-class, non-white residents and thereby operates as an instrument of settler-colonialism, racial-capitalism, and neoliberalism. From their positions as students, immigrants, and activists, they argue that Pittsburgh offers an illustrative story: the luring of the creative class turns universities and their professors into engines that drive and legitimate the speculative demand for "smart" ventures that unfold as neoliberal projects of public space extractivism. Likewise in the United States, media scholar and artist Megan Wiessner's photo essay traces mass timber as smart material in the Pacific Northwest. Her images illustrate how industry discourses of digital workflow and precision fabrication are deeply rooted in the settler colonial terrain of logging, dispossession, and racial exclusion. Here, smartness is not merely a matter of digital workflows and precision fabrication but deeply entangled in the dispossession of Indigenous land and labour struggle. In the final contribution of this section, media scholar June Yeoreum Kim brings us to South Korea and Vietnam, where smart cities reveal a long legacy of land clearing that has become an essential foundation for fostering smartness. Kim demonstrates how computer ideas of deleting, formatting, and uploading are rebooting processes that are deeply inscribed in the materiality of land itself.

"Plastering Frictions and Fissures with Smartness" is the second part, which takes up the anthropological idiom of friction that occurs when universals, that is, a smart scheme, policy, or plan, "hit the ground" of the particular (Tsing 2011) while disguising other social, political, or environmental frictions and fissures. Contributors show how influential actors attempt to smooth out political and ethical frictions to create the illusion of whole parts in networks void of humans. In chapter 8, anthropologist Junnan Mu takes us on an ethnographic journey below the smooth surface of Kenya's first smart city. Here, incoherent meanings manifest on the ground in what Mu calls "definitional impasse": a yearning for a postcolonial urban future that inheres a vacuum between an abstract, imagined ideal and fundamental material realities. The "smart city" in an African locale is the result of multiple, often contradictory factors: local socioeconomic transformation, individual aspirations, and global speculations of digital futures. Even (and perhaps especially) when the term "smart city/system" is not evoked, the definitional impasse offers a valuable concept to analyse technoscientific smartness more generally and extends scholarship on "smartness" as empty signifier (Burns et al. 2023). Chapter 9, by Gabriel Dorthe and Laure Dobigny, discusses another "universalist concept," the smart building, at a French university campus. As embedded ethnographers in a research project, their account traces frictions that emerged when engineers attempted to develop obedient yet autonomous smart users. Consequently, what counted as "smart" was the pragmatic adaptation of technical systems, such as stickers nudging residents to switch off their lights or a door stopper regulating the otherwise automated cooling system. Architectural researcher Simon Rabyniuk next shows that in "vertiplaces" the scaling of commercial drones from test bed to national systems may break due to so-called thin simplifications of local variations. His contribution points to the near future where smartness may be, literally, made out of (thin) air. Sarah Marquis rounds off this section with a discourse analysis of "climate-smart" agriculture in the Canadian context. Her analysis shows how proponents of such smart farming systems that use variable rate application – which allow farmers to use spatial data to measure productivity in their fields – often use "sustainable" and "smart" interchangeably to the extent that the latter may replace the former (see also Gabrys 2016; Gugganig forthcoming). Yet such seemingly effective association with the empty signifier "smart" cannot disguise the fact that smartness as sustainable optimization ultimately fails to repair, and rather plasters over, a broken agricultural system.

The third part, "Smartification as Boundary Work," resonates with a classic STS-approach, as contributions inquire into various ways in which actors create, advocate, and contest boundaries (Gieryn 1995) around smart spaces, objects, and people (see Klimburg-Witjes 2021). In chapter 13, architecture scholar Ali Fard asks where to locate the smart city when one follows the material conditions of its palimpsest, "the cloud," into faraway data centres, extraction sites, and dumping grounds. Smartification processes can entail a constant boundary-drawing around the smart city "as a wholly knowable, manageable, and ultimately controllable archipelago in a sea of an undefined periphery" (139). Critical analyses like Fard's uncover the porosity of the city and of smartness, and they clearly delineate the reliant global material geographies (see also

Unknown Fields Division 2015). The next contribution, a Visual Vignette by STS scholar Sebastian Bornschlegl, re-presents a smart city project in Vienna, Austria, that is split into multiple smaller, partially connected oases of smartness. This partiality is also reflected in the author's artistic mode of collaging his Visual Vignette, a photo essay that transgresses the boundary between image and text (Gugganig and Douglas-Jones 2021). In chapter 15, anthropologist Hannah Carlan ethnographically explores the boundary-making of a smart city scheme in Dharamshala, India. Different from most smart cities' aim of fencing off undesirable people, animals, or places, Dharamshala's smart city, she highlights, attempts to extract such undesirables from *within* the physical boundaries of the smart city, as if they have been trapped in order to be digested – similar to their smart bins – by the logic, means, and overall processes of smartification. The rendering of living and non-living entities as disposable (and eventually absorbable), Carlan argues, is not a by-product of smartness, but its very raison d'être. Finally, poet Devin Shepherd and artist Juliette Walker present their reprinted zine *Wal*Smartification: Considering the Superimposition of Dockless Shared Electric Scooters on Fayetteville,* which evocatively visualizes smart scooters in Arkansas (US) as urban clutter, and thus a tale of neoliberal "Walsmarting" or "free-market smartness" of degraded public space (184). Through its undemocratic logic, the ability to use e-scooters (that is, deciphering the e-scooter system and signs, having access to a cell phone, etc.) draws boundaries around digitally literate "smart" citizens and spaces, while nebulous regulations of where one can drive and park permits them to clutter public space of not-as-smart pedestrians. As they explain, "for the smartest to remain the smartest means preventing others from getting access to information, which maintains the relative scarcity of smartness" (Shepherd and Walker 181, personal conversation).

Contributions to part 4, "Opening up Smartness," are akin to another common STS move of "opening up" the black box of technical systems to study their inherent social relations (Pinch 1986), that is, smartness attributions to the urban (Tironi and Sánchez Criado 2015) and rural contexts. In chapter 18, STS scholar Martin Abbott explores the US-federal smart flood technology "Risk Rating 2.0 – Equity in Action" in New Orleans, whose algorithm merely considers economic fairness. Opening up the marketing campaign "FloodSmart," Abbott reveals that it is more about equity inaction and the associated long-held, unaddressed, racial and environmental injustices in the region. With the case of AI-assisted traffic lights in Vienna, Austria, Poya Sepehr then demonstrates how smartification processes could also open up opportunities for mobility justice; jaywalking – the label given to people disregarding traffic light rules and a reference to the jay bird often connoted as "dumb" – may turn into a self-empowering, smart act of navigating an unjust, car-dominated mobility system, with its traffic lights serving as devices of (social) control. Next, policy researcher S. Ashleigh Weeden offers an exploratory framework of rural smartification. Her critique is twofold and turns towards the measure of rural areas against the city, which always have to "catch up" with urban standards. She shows how the common urban bias in critical analyses of smart systems is also evident in rural relations, where smartification gets collapsed with urbanization. The section ends with media scholar Jean Hardy's chapter, "Rural Expertise and the Sewer," in which he analyses the introduction of digital asset management schemes in rural Michigan (US). His ethnographic account opens up universal assumptions of "smart" systems – here, to improve sewage systems regardless of a specific place – to show how they butt up against embodied, site-specific sewer expertise. Digital systems, he concludes, require much more sensitive integration into existing labour relations.

The final section, "Rethinking Smartness," offers a variety of mostly artistic contributions that explore how smartification processes may be rethought – and re-experienced – in alternative readings of smartness. What unites the contributions is the move beyond simplistic technoscientific conceptions of smartness to consider smartness as historically and culturally grounded, more-than-human intelligence that is emergent in social relations. Situated in Pakistan, in chapter 23 architect Soha Macktoom and anthropologist Aqdas Fatima offer an analysis of Karachi where western notions of modernity in "world class city" aesthetics are imposed onto this postcolonial city. Their analysis entails three evocative art panels (fig. 23.1), which illustrate the juxtaposition of inefficient "smart" cooling systems, such as air conditioning units on glass-cladded high-rise buildings, with autochthonous "unsmart" cooling practices like the use of fabric for shade or clay pots for storing drinking water (fig. 23.7, which also serves as this book's cover image). Their use of the term "unsmart" resembles scholarship on idiotic smartness (Tironi and Valderrama 2019) to suggest locally embedded and historically grown modes of living that resist dominant notions of technoscientific smartness. Artist and researcher Bruno Moreschi and political scientist Gabriel Pereira then present their short film *Future Movement Future – REJECTED*, where a panel of academics reviews and ends up rejecting a proposal for using an algorithmic model through smart cameras to predict the movement of cars. Their rejection is a cautionary science fiction tale that contributes to creating what Moreschi and Pereira call a "Positive History of NO" to smartness (266). Their provocation invites reflections not only on what kind of smartness is aspired/rejected but also on how and by whom these decisions ought to be made. Next, in chapter 25 artist and researcher Carola Moujan takes the reader on a journey into multispecies entanglements in the "smart city," which critique and expand this generic concept, containing trees as intelligent cohabitants. With her artwork, Moujan explores new, "eco-logical," non-anthropocentric methodologies and principles that move beyond binaries between digital and biological intelligence to instead consider their relationalities in what she calls a "*Forest City* vision of urban smartness" (272). The section and edited volume ends with a short film by mathematician Claire Glanois and artist and designer Vytautas Jankauskas. In *Unfamiliar Convenient*, a voice assistant and spiritual vacuum cleaner invite the viewer to consider our domestic space as an experimental site, where we form curious relationships with smart devices. Their work reminds us of the *Technological Dreams Series* by designers Robert Dunne and Fiona Raby, which addresses how we (will) live with technological cohabitants, like robots, when they have distinct personalities.[11] These relationships may be better understood as continuous, situated renegotiations of what it means to be smart in contemporary times and places.

Where to Go from Here?

There are certain defiant factors and virtues that remain outside of the control of smart systems, like the smart city, as they are not (controllable) computers (Mattern 2021). This has turned scholars' and activists' attention – again, in the urban context – to alternative visions and concrete efforts where technology is not the driving force behind smart solutions, but rather collective ideas and action, communal resilience, and locally grounded approaches to innovation (Cardullo et al. 2019; de Lange and de Waal 2013; Hollands 2016; Joss et al. 2019, 24). As succinctly pointed out by Karrie Jacobs:

> The real problem is that with their emphasis on the optimization of everything, smart cities seem designed to eradicate the very thing that makes cities wonderful. New York and Rome and Cairo (and Toronto) are not great cities because they're efficient: people are attracted to the messiness, to the compelling and serendipitous interactions within a wildly diverse mix of people living in close proximity. (2022)

Such defiance may counter optimization, efficiency, and interconnectivity because they do not reflect virtues of human and other species, like trees,

---

11 See https://dunneandraby.co.uk/content/projects/10/0#, accessed June 5, 2024.

dogs, or air. It raises the question: "What if the smart city were one that promoted human and non-human flourishing?" (Burns et al. 2023, 20).

Some scholars have therefore put into question whether it is even possible to retain smartness as a guiding principle when pursuing justice, considering instead what a "dumb city" that rejects digital technologies could look like (Burns et al. 2023). In consequence, the "idiotic," non-conforming citizen can turn into a generative mode of divergence and rejection of the smart city (Tironi and Valderrama 2019). In the words of Jennifer Gabrys, the figure of the idiot is not insulting, in the sense of a "dumb" person, but "someone or something that causes us to think about and encounter the complexities of participation and social life as something other than prescribed or settled (2016, 209). In this volume, Abhishek Viswanathan and Bobbie Fan offer their personal experiences as students and organizers in Pittsburgh to consider how techno-solutionist motivations of academia standardize cities as urban laboratories that by default devalue communities and their relational smartness. Within and beyond cities, Kintzi and Faxon and Macktoom and Fatima show that the frictions in lived realities of those having to adapt to so-called smart technologies have the potential to reshape ways in which smartification unfolds. Their collective awareness is more grounded than technoscientific smartness, whether it is in the need for better public infrastructure and successful community-led proposals to halt self-driving public transit and predictive policing (Viswanathan and Fan, this volume), fostered resident-researcher solidarity, or the implementation of solutions to people's real-life problems including better irrigation, affordable energy, or well-adapted cooling systems (as we'll see in other contributions).

What has also emerged in critical discussions is a move in the other direction, calling for an appropriation of the term "smart" – similar to how critical actors have claimed "queer" or "innovation" for their purposes. Who gets to define what is "smart" (Hollands 2008), and what would an emancipatory smartness across sectors, spaces, objects, and living organisms (humans, other animals, etc.) look like? One way of engaging in such emancipatory work is to occupy smartness through bottom-up approaches. Offering an overview of DIY, grassroots efforts along with or counter to smart city initiatives, Tironi and Sánchez Criado argue that a rearticulation of smartness not only means generating different forms of data but also the articulation of "open-source infrastructures that redistribute *smartness*" among a variety of actors (2015, 90; emphasis original). A less urban-centric example in that regard is the German Smart Coop,[12] which refers to cooperatives that have existed

---

12 https://smartde.coop/, accessed June 2, 2024.

for decades, if not centuries, and make use of (non-commercial) platforms to network with similar initiatives. Yet one should remain cautious to think that adapting either of such "radical" approaches – dismissing or claiming smartness – is the solution. As Mu convincingly shows in her ethnographic account in this volume, for many Kenyans it is crucial to adopt flexible and adaptive approaches to and identifications with the "smart city" while they "shape their own identities and envision their future in a postcolonial society" (91–2). In other words, binary approaches – once again and still – only serve the illusion of modernism (Latour 1993) while dismissing postcolonial realities and the need for nuanced, empirically grounded accounts.

In regards to rural relations and agriculture, Kelly Bronson argues that smart farming requires a fundamental rethinking of innovation, for instance, in the form of locally adapted cropping strategies not necessarily dependent on technical sensors. The algorithm, she argues, is "a messy, fragile, and inherently political accomplishment, [and] life is so dazzlingly complex it has quiet corners of resistance that allude [to] any form of knowledge creation" (2022, 152). Indeed, as Weeden shows in this volume, rural places can be understood as sites of conceptual and material resistance to neoliberal smartification agendas, which often reflect urban-biased "smart" policies and development schemes. Critical scholarship on smartness is also moving beyond the urban bias; however, the rural ought not to be understood as novel intellectual extraction site. Rather, scholars should move beyond simplistic *binary* conceptions of smartness through analyses across these geographic spheres. Consequently, the "right to be rural" movement would make important interventions in the dominant "digital divide" paradigm (Burns et al. 2023; Masucci et al. 2020), again, between the urban and rural, where rural broadband infrastructures are seen as inevitable. What could instead be developed are capacities and nuanced "relationships with technology that serve self-determined, self-governed rural interests" (Weeden, this volume, 225).

So, in order not to fall into the trap of perpetuating binaries between the urban and the rural, the technical and the "natural," between digital and biological intelligence (Moujan, this volume), it is crucial to highlight instances in which humans and other species interact with technically smart devices in surprising, intimate ways, as Jankauskas and Glanois's short film demonstrates. How do we make sense of social relations with smart home gadgets that become like pets, even children, and that may become humans' kin? This is not fundamentally new stuff, as decades ago Donna Haraway (1991) challenged our thinking to consider ourselves as cyborgs: as always already half-human, half-machine. What is perhaps new is that humans are reconsidering what it means to be "smart," both in terms of a deeper consideration of what is human about it, more-than-human, and in how they themselves relate to technoscientific smartness.

How we study such diversity of smartness also reflects its analytic diversity: just as much as scholars and artists find that "empty" smart systems, spaces, objects, and people are "filled" with meaning according to specific sociocultural, political, and economic contexts, studying said diversity can take shape in a variety of approaches. Attending to smartification processes as "an unfolding, contradictory, and deeply contested arena of politics" (Kintzi and Faxon, this volume, 41) – more specifically, to actors' practices and performances of envisioning, manifesting, and contesting smartness – can help to complicate the too-often-made simplistic equations of "smart" with technoscientific and technocratic.

References

Araya, Daniel, ed. 2015. *Smart Cities as Democratic Ecologies.* Palgrave Macmillan. https://doi.org/10.1057/9781137377203.

Arboleda, Martín. 2020. *Planetary Mine: Territories of Extraction under Late Capitalism.* Verso.

Arens, Pit, and Martina Schlünder. 2020. "Panels and Frames: Toward a New Relationship between Text and Image in Academic Writing." In *Boxes: A Field Guide,* edited by Susanne Bauer, Martina Schlünder, and Maria Rentetzi. Mattering Press. http://doi.org/10.28938/9781912729012.

Ashby, Madeline, Farhad Pakdel, and sava saheli singh. 2019. "Frames, 11 min." www.surveillance-studies.ca/projects/screening-surveillance/frames.

Aurigi, Alessandro, and Nancy Odendaal. 2020. "From 'Smart in the Box' to 'Smart in the City': Rethinking the Socially Sustainable Smart City in Context." *Journal of Urban Technology* 28 (1–2): 55–70. https://doi.org/10.1080/10630732.2019.1704203.

Benessia, A., and Â.G. Pereira, eds. 2015. "The Dream of the Internet of Things: Do We Really Want and Need to Be Smart?" In *Science, Philosophy and Sustainability: The End of the Cartesian Dream.* Routledge.

Borgdorff, Henk, Peter Peters, and Trevor Pinch, eds. 2020. "Dialogues Between Artistic Research and Science and Technology Studies: An Introduction." In *Dialogues between Artistic Research and Science and Technology Studies.* Routledge. https://doi.org/10.4324/9780429438875-1.

Bowker, Geoffrey C. 2013. "Data Flakes: An Afterword to *'Raw Data' Is an Oxymoron.*" In *"Raw Data" Is an Oxymoron,* edited by Lisa Gitelman. MIT Press. https://doi.org/10.7551/mitpress/9302.003.0011.

Brenner, Neil, and Christian Schmid. 2011. "Planetary Urbanisation." In *Urban Constellations,* edited by Matthew Gandy. JOVIS.

Bronson, Kelly. 2019. "Looking Through a Responsible Innovation Lens at Uneven Engagements with Digital Farming." *NJAS-Wageningen Journal of Life Sciences* 90–1 (1): 1–6. https://doi.org/10.1016/j.njas.2019.03.001.

– 2022. *The Immaculate Conception of Data: Agribusiness, Activists, and Their Shared Politics of the Future*. McGill-Queen's Press. https://doi.org/10.1515/9780228012535.

Bronson, Kelly, and Irena Knezevic. 2016. "Big Data in Food and Agriculture." *Big Data & Society* 3 (1): 1–5. https://doi.org/10.1177/2053951716648174.

Burns, Ryan, Victoria Fast, and Debra Mackinnon. 2023. "Introduction: Towards Urban Digital Justice: The Smart City as an Empty Signifier." In *Digital (In)justice in the Smart City*, edited by Debra Mackinnon, Ryan Burns, and Victoria Fast. University of Toronto Press.

Burrell, Jenna. 2020. "On Half-Built Assemblages: Waiting for a Data Center in Prineville, Oregon." *Engaging Science, Technology, and Society* 6 (June): 283–305. https://doi.org/10.17351/ests2020.447.

Cardullo, Paolo, Cesare Di Feliciantonio, and Rob Kitchin. 2019. *The Right to the Smart City*. Emerald. https://doi.org/10.1108/9781787691391.

Chun, Wendy H.K. 2011. "Crisis Crisis Crisis, or Sovereignty and Networks." *Theory, Culture & Society* 28 (6): 91–112. https://doi.org/10.1177/0263276411418490.

Cronon, William. 1991. *Nature's Metropolis: Chicago and the Great West, 1848–1893*. W.W. Norton.

Datta, Ayona. 2015. "New Urban Utopias of Postcolonial India: 'Entrepreneurial Urbanization' in Dholera Smart City, Gujarat." *Dialogues in Human Geography* 5 (1): 3–22. https://doi.org/10.1177/2043820614565748.

– 2023. "Complicated and Complicating Digital Divides: A Dialogue with Ayona Datta." In *Digital (In)Justice in the Smart City*, edited by Debra Mackinnon, Ryan Burns, and Victoria Fast. University of Toronto Press.

Datta, Ayona, and Nancy Odendaal. 2019. "Smart Cities and the Banality of Power." *Environment and Planning D: Society and Space* 37 (3): 387–92. https://doi.org/10.1177/0263775819841765.

de Lange, Michiel, and Martijn de Waal. 2013. "Owning the City: New Media and Citizen Engagement in Urban Design." *First Monday* 18 (11). http://firstmonday.org/ojs/index.php/fm/article/view/4954/3786.

Elden, Stuart, and Adam David Morton. 2016. "Thinking Past Henri Lefebvre: Introducing 'The Theory of Ground Rent and Rural Sociology.'" *Antipode* 48 (1): 57–66. https://doi.org/10.1111/anti.12171.

Favero, Paolo. 2017. "Tainted Frictions: A Visual Essay." *American Anthropologist* 119 (2): 361–64. https://doi.org/10.1111/aman.12871.

Felt, Ulrike. 2025. *Academic Times: Contesting the Chronopolitics of Research*. Palgrave Macmillan Singapore. https://doi.org/10.1007/978-981-96-4609-8.

Fischer, Michael M. 2003. *Emergent Forms of Life and the Anthropological Voice.* Duke University Press. https://doi.org/10.1515/9780822384953.
Foth, M., M. Brynskov, and T. Ojala. 2015. *Citizen's Right to the Digital City.* Springer. https://doi.org/10.1007/978-981-287-919-6.
Fraser, Alistair. 2019. "Land Grab/Data Grab: Precision Agriculture and Its New Horizons." *The Journal of Peasant Studies* 46 (5): 893–912. https://doi.org/10.1080/03066150.2017.1415887.
Gabrys, Jennifer. 2014. "Programming Environments: Environmentality and Citizen Sensing in the Smart City." *Environment and Planning D: Society and Space* 32 (1): 30–48. https://doi.org/10.1068/d16812.
– 2016. *Program Earth: Environmental Sensing Technology and the Making of a Computational Planet.* University of Minnesota Press. https://doi.org/10.5749/minnesota/9780816693122.001.0001.
– 2020. "Smart Forests and Data Practices: From the Internet of Trees to Planetary Governance." *Big Data & Society* 7 (1): 1–10. https://doi.org/10.1177/2053951720904871.
– 2022. "Programming Nature as Infrastructure in the Smart Forest City." *Journal of Urban Technology* 29 (1): 13–19. https://doi.org/10.1080/10630732.2021.2004067.
Galle, Nadina J., Sophie A. Nitoslawski, and Francesco Pilla. 2019. "The Internet of Nature: How Taking Nature Online Can Shape Urban Ecosystems." *The Anthropocene Review* 6 (3): 279–87. https://doi.org/10.1177/2053019619877103.
Gieryn, Thomas F. 1995. "Boundaries of Science." In *Handbook of Science and Technology Studies*, edited by Sheila Jasanoff, Gerald E. Markle, James C. Petersen, and Trevor Pinch. Sage Publications. https://doi.org/10.4135/9781412990127.n18.
Greenfield, Adam. 2013. *Against the Smart City: A Pamphlet. This is Part I of "The City Is Here to Use."* Do Projects.
Gugganig, Mascha. 2025. "Vanguard Visions of Vertical Farming: Envisaging and Contesting an Emerging Food Production System." *Science, Technology, & Human Values.* https://doi.org/10.1177/01622439241240796.
– Forthcoming. "Fixing Sustainability Through Technoscience, and the Politics of Diversity: The Case of EU Agriculture." *Environmental Science & Policy.*
Gugganig, Mascha, and Kelly Bronson. 2022. "Digital Agriculture and the Promise of Immateriality." In *Food Studies: Matter, Meaning, Movement*, edited by David Szanto, Amanda Di Battista, and Irena Knezevic. Food Studies Press.
Gugganig, Mascha, and Rachel Douglas-Jones. 2021. "Visual Vignettes." In *Sensing In/Security: Sensors as Transnational Security Infrastructures*, edited by Nina Klimburg-Witjes, Nikolaus Poechhacker, and Geoffrey C. Bowker. Mattering Press. http://doi.org/10.28938/9781912729111.

Günel, Gökçe. 2019. *Spaceship in the Desert: Energy, Climate Change, and Urban Design in Abu Dhabi*. Durham: Duke University Press. https://doi.org/10.1215/9781478002406.

Halpern, Orit, and Gökçe Günel. 2017. "FCJ-215 Demoing unto Death: Smart Cities, Environment, and Preemptive Hope." *The Fibreculture Journal* 29: 1–23. https://doi.org/10.15307/fcj.29.215.2017.

Halpern, Orit, and Robert Mitchell. 2023. *The Smartness Mandate*. MIT Press. https://doi.org/10.7551/mitpress/14623.001.0001.

Halpern, Orit, Robert Mitchell, and Bernard Dionysius Geoghegan. 2017. "The Smartness Mandate: Notes toward a Critique." *Grey Room* 68: 106–29. https://doi.org/10.1162/GREY_a_00221.

Haraway, Donna. 1991. "A Cyborg Manifesto: Science, Technology, and Socialist-Feminism in the Late Twentieth Century." In *Simians, Cyborgs and Women: The Reinvention of Nature*, 149–81. Routledge.

— 2016. "Sympoiesis: Symbiogenesis and the Lively Arts of Staying with the Trouble." In *Staying with the Trouble: Making Kin in the Chthulucene*. Duke University Press. https://doi.org/10.2307/j.ctv11cw25q.

Helmreich, Stefan, and Caroline A. Jones. 2018. "Science/Art/Culture through an Oceanic Lens." *Annual Review of Anthropology* 47: 97–115. https://doi.org/10.1146/annurev-anthro-102317-050147.

Hilgartner, Stephen. 2015. "Capturing the Imaginary: Vanguards, Visions and the Synthetic Biology Revolution." In *Science and Democracy*, edited by Stephen Hilgartner, Clark Miller, and Rob Hagendijk. Routledge. https://doi.org/10.4324/9780203564370.

Hilgartner, Stephen, and Charles L. Bosk. 1988. "The Rise and Fall of Social Problems: A Public Arenas Model." *American Journal of Sociology* 94 (1): 53–78. https://doi.org/10.1086/228951.

Hogan, Mél. 2018. "Big Data Ecologies." *Ephemera* 18 (3): 631–57.

Hollands, Robert G. 2008. "Beyond the Corporate Smart City? Glimpses of Other Possibilities of Smartness." In *Smart Urbanism: Utopian Vision or False Dawn?* edited by Simon Marvin, Andrés Luque-Ayala, and Colin McFarlane. Routledge. https://doi.org/10.4324/9781315730554.

Jacobs, Karrie. 2022. "Toronto Wants to Kill the Smart City Forever." *MIT Technology Review* (July/August). www.technologyreview.com/2022/06/29/1054005/toronto-kill-the-smart-city/.

Jones, Caroline A. 2005. "Doubt Fear." *Art Papers* 29 (1): 24–35.

— 2021. "Foreword by Caroline A. Jones." In *Routledge Handbook of Art, Science, and Technology Studies*, edited by Hannah Star Rogers, Megan Halpern, Dehlia Hannah, and Kathryn de Ridder-Vignone. Routledge.

Joss, Simon, Frans Sengers, Daan Schraven, Federico Caprotti, and Youri Dayot. 2019. "The Smart City as Global Discourse: Storylines and Critical Junctures across 27 Cities." *Journal of Urban Technology* 26 (1): 3–34. https://doi.org/10.1080/10630732.2018.1558387.

Karvonen, Andrew, Federico Cugurullo, and Federico Caprotti, eds. 2019. *Inside Smart Cities: Place, Politics and Urban Innovation*. Routledge. https://doi.org/10.4324/9781351166201.

Kehm, Walter H. 2020. *Accidental Wilderness: The Origins and Ecology of Toronto's Tommy Thompson Park*. University of Toronto Press. https://doi.org/10.3138/9781487538040.

Kitchin, Rob. 2011. "The Programmable City." *Environment and Planning B: Planning and Design* 38 (6): 945–51. https://doi.org/10.1068/b3806com.

– 2014. "The Real-Time City? Big Data and Smart Urbanism." *GeoJournal* 79 (November): 1–14. https://doi.org/10.1007/s10708-013-9516-8.

Klimburg-Witjes, Nina, Nikolaus Poechhacker, and Geoffrey C. Bowker. 2021. *Sensing In/Security: Sensors as Transnational Security Infrastructures*. Mattering Press. https://doi.org/10.28938/9781912729111.

Kumar, Ankit. 2019. "Beyond Technical Smartness: Rethinking the Development and Implementation of Sociotechnical Smart Grids in India." *Energy Research & Social Science* 49 (March): 158–68. https://doi.org/10.1016/j.erss.2018.10.026.

Laclau, Ernesto. 2007. *Emancipation(s)*. Verso.

Latour, Bruno. 1993. *We Have Never Been Modern*. Harvard University Press.

Latour, Bruno, and Peter Weibel, eds. 2020. *Critical Zones: The Science and Politics of Landing on Earth*. MIT Press.

Lefebvre, Henry. (1973) 1976. *The Survival of Capitalism: Reproduction of the Relations of Production*. Allison and Busby.

Levenda, Anthony M., and Dillon Mahmoudi. 2019. "Silicon Forest and Server Farms: The (Urban) Nature of Digital Capitalism in the Pacific Northwest." *UMBC Geography and Environmental Systems Department Collection*: 1–14.

Li, Tania Murray. 2007. *The Will to Improve: Governmentality, Development, and the Practice of Politics*. Duke University Press.

Luque-Ayala, A., C. McFarlane, and S. Marvin. 2016. "Introduction." In *Smart Urbanism: Utopian Vision or False Dawn?* edited by Simon Marvin, Andrés Luque-Ayala, and Colin McFarlane. Routledge.

Maschewski, Felix, and Anna-Verena Nosthoff, eds. 2022. "Big Tech and the Smartification of Agriculture: A Critical Perspective." In *The State of Big Tech 2022, IT for Change*. https://ssrn.com/abstract=4080210.

Masucci, Michele, Hamil Pearsall, and Alan Wiig. 2020. "The Smart City Conundrum for Social Justice: Youth Perspectives on Digital Technologies and Urban Transformations." *Annals of the American Association of Geographers* 110 (2): 476–84. https://doi.org/10.1080/24694452.2019.1617101.

Mattern, Shannon. 2021. *A City Is Not a Computer: Other Urban Intelligences*. Princeton University Press. https://doi.org/10.1515/9780691226750.

McFarlane, Colin, and Ola Söderström. 2017. "On Alternative Smart Cities: From a Technology-Intensive to a Knowledge-Intensive Smart Urbanism." *City* 21 (3–4): 312–28. https://doi.org/10.1080/13604813.2017.1327166.

Miles, Christopher. 2019. "The Combine Will Tell the Truth: On Precision Agriculture and Algorithmic Rationality." *Big Data & Society* 6 (1): 1–12. https://doi.org/10.1177/2053951719849444.

Misrach, Richard, and Kate Orff. 2012. *Petrochemical America*. Aperture.

Negroponte, Nicholas. 1970. *The Architecture Machine: Toward a More Human Environment*. The MIT Press. https://doi.org/10.7551/mitpress/8269.001.0001.

Palanca-Castan, Nicolas, Beatriz S. Tajadura, and Rodrigo Cofré. 2021. "Towards an Interdisciplinary Framework about Intelligence." *Heliyon* 7 (2): e06268. https://doi.org/10.1016/j.heliyon.2021.e06268.

Pinch, Trevor J. 1986. *Confronting Nature: The Sociology of Solar-Neutrino Detection*. Kluwer Academic Publishers.

— 2021. "Foreword by Trevor Pinch." In *Routledge Handbook of Art, Science, and Technology Studies*, edited by Hannah S. Rogers, Megan K. Halpern, Dehlia Hannah, and de Kathryn Ridder-Vignone. Routledge.

Rajan, Kaushik S. 2006. *Biocapital: The Constitution of Postgenomic Life*. Duke University Press. https://doi.org/10.2307/j.ctv120qqqr.

Rogers, Hannah S. 2022. *Art, Science, and the Politics of Knowledge*. MIT Press. https://doi.org/10.7551/mitpress/13885.001.0001.

Rogers, Hannah S., Megan K. Halpern, Dehlia Hannah, and Kathryn de Ridder-Vignone, eds. 2021. *Routledge Handbook of Art, Science, and Technology Studies*. Routledge. https://doi.org/10.4324/9780429437069.

Rosol, Marit, and Gwendolyn Blue. 2022. "From the Smart City to Urban Justice in a Digital Age." *City* 26, no. 4: 684–705. https://doi.org/10.1080/13604813.2022.2079881.

Ross, Andrew. 1993. "The New Smartness." *Science as Culture* 4 (1): 94–109. https://doi.org/10.1080/09505439309526375.

Sadowski, Jathan. 2020. *Too Smart: How Digital Capitalism Is Extracting Data, Controlling Our Lives, and Taking Over the World*. MIT Press. https://doi.org/10.7551/mitpress/12240.001.0001.

Sadowski, Jathan, and Roy Bendor. 2019. "Selling Smartness: Corporate Narratives and the Smart City as a Sociotechnical Imaginary." *Science, Technology, & Human Values* 44 (3): 540–63. https://doi.org/10.1177/0162243918806061.

Salter, Chris. 2022. "Utopia Unresolved." *MIT Technology Review* 125 (4): 66–70. https://www.technologyreview.com/2022/06/24/1053969/smart-city-unrealized-utopia/.

Scott, James C. 2008. "Taming Nature: An Agriculture of Legibility and Simplicity." In *Seeing Like a State: How Certain Schemes to Improve the Human Condition Have Failed*, 262–306. Yale University Press.

Seiner, Tanja. 2016. "Touch Deeper." Exhibition Catalogue, Lothringer13 Gallery, Munich. Accessed September 1, 2023. https://tanjaseiner.de/.

Sheikh, Hira, Peta Mitchell, and Marcus Foth. 2023. "More-than-Human Smart Urban Governance: A Research Agenda." *Digital Geography and Society* 4: 1–13. https://doi.org/10.1016/j.diggeo.2022.100045.

Slayton, Rebecca. 2013. "Efficient, Secure Green: Digital Utopianism and the Challenge of Making the Electrical Grid 'Smart.'" *Information & Culture* 48 (4): 448–78. https://doi.org/10.7560/IC48403.

Strathern, Marilyn, ed. 2000. *Audit Cultures: Anthropological Studies in Accountability, Ethics, and the Academy.* Routledge.

Tironi, Martín, and Matías Valderrama. 2019. "Acknowledging the Idiot in the Smart City: Experimentation and Citizenship in the Making of a Low-Carbon District in Santiago de Chile." In *Inside Smart Cities: Place, Politics and Urban Innovation*, edited by Andrew Karvonen, Federico Cugurullo, and Federico Caprotti. Routledge. https://doi.org/10.4324/9781351166201-11.

Tironi, Martin, and T. Sánchez Criado. 2015. "Of Sensors and Sensitivities: Towards a Cosmopolitics of 'Smart Cities'?" *Tecnoscienzia* 6 (1): 89–108. https://doi.org/10.6092/issn.2038-3460/17240.

Tsing, Anna L. 2011. *Friction: An Ethnography of Global Connection.* Princeton University Press. https://doi.org/10.2307/j.ctt7s1xk.

Unknown Fields Division. 2015. "Rare Earthenware." Film, 7 min. https://vimeo.com/124621603.

Vanolo, Alberto. 2014. "Smartmentality: The Smart City as Disciplinary Strategy." *Urban Studies* 51 (5): 883–98. https://doi.org/10.1177/0042098013494427.

Watson, Vanessa. 2013. "African Urban Fantasies: Dreams or Nightmares?" *Environment & Urbanization* 26 (1): 215–31. https://doi.org/10.1177/0956247813513705.

Weeden, Sara Ashleigh. 2022. "The Right to Multiple Futures in the Shadow of Canada's Smart City Movement." In *The Right to be Rural*, edited by Karen Foster and Jennifer Jarman, 253–70. University of Alberta Press. https://doi.org/10.1515/9781772125955-016.

Yigitcanlar, Tan, Marcus Foth, and Md. Kamruzzaman. 2019. "Towards Post-Anthropocentric Cities: Reconceptualizing Smart Cities to Evade Urban Ecocide." *Journal of Urban Technology* 26 (2): 147–52. https://doi.org/10.1080/10630732.2018.1524249.

Zhao, Hailing, and Rachel Douglas-Jones. 2022. "Weaving the Net: Making a Smart City through Data Workers in Shenzhen." *East Asian Science, Technology and Society: An International Journal* 16 (4): 461–85. https://doi.org/10.1080/18752160.2022.2088919.

PART ONE

The (Settler) Colonial Roots of Smartification

PART ONE

2 Introduction to the (Settler) Colonial Roots of Smartification

MASCHA GUGGANIG

It might be intuitive to start with sites of the Global North that are equivalent to where the technoscientific rendering of "smartness" was first envisioned. While some argue that the term was first coined by IBM "in hopes that technology could improve the way cities functioned," Jacobs (2022) notes that as a "strategy for city-building, it's been most successfully deployed under authoritarian regimes (Putin is a fan)." Beyond authoritarian regimes, the manifestation of smart systems stretches historically deep and geographically wide. More specifically, smartification processes are deeply rooted in developmentalist, postcolonial, and settler colonial structures, and as such they cut through common (binary) analytical categories, such as the urban and the rural, the Global North and the Global South. Because of this, articulations and manifestations of smartness are deeply entangled in fractured, often widely dispersed colonial conditions, and studying their spatial dimensions offers important insights into the tensions between globalization, local profiteers, and efforts of self-determination. Smartification processes often operate to disguise settler colonial land extraction. This is evident when the prefix "smart" is merely added to any progressivist project or political agenda aiming at economic development. Yet the common "smooth visions" of smartness simply replicate existing inequalities and reveal an (equally common) extractivist logic, such as grabbing land and material sources. What is new, as Kintzi and Faxon show in their chapter, is the framing of digitally networked infrastructures and subjects as the solution to all kinds of economic, social, and environmental ailments. Smartness, such as in "climate-smart forestry" (Wiessner), now may include datafication and climate change, so long as they are rendered (manageable) problems.

Part 1, "The (Settler) Colonial Roots of Smartification," demonstrates these dynamics in a variety of places, including Jordan, Myanmar, the

United States, South Korea, and Vietnam. In relation to Myanmar's smart farms and Jordan's smart grid system, Kendra Kintzi and Hilary Faxon demonstrate that just as with "development," the appendix of "smart" has turned into an industry buzzword where technologies will fix any social and economic (agricultural and energy) problems. Abhishek Viswanathan and Bobbie Fan then turn to academic smartification in Pittsburgh's (US) nexus of universities, foundations, corporations, and politicians that advances a "Creative City" and smart (read: surveillance) agenda. This agenda functions as an instrument of settler colonialism, racial-capitalism, and neoliberalism by keeping out working-class, non-white residents, effectively rendering them as anything but smart. As students, immigrants, and activists, the authors demonstrate Pittsburgh as an illustrative case for how the creative class turns universities into engines of "smart" ventures, unmasking them as neoliberal public space extractivist projects. The two contributions also show that alternative approaches to smartification can be envisioned and manifested when grounded in people's lived experiences – such as Myanmar farmers calling for better irrigation, or Pittsburgh's residents testifying on labour impacts of autonomous tech epistemology, using thick blankets and rugs to keep the cold out, and students organizing for grassroots interventions.

On the Pacific Northwest Coast of the United States, Megan Wiessner's photo essay evocatively visualizes smartification as it happens in mass timber fabrication. In her photos, climate-smart forestry encompasses digital workflows and precision fabrication that follow a long history of settler colonialism in the form of logging, dispossession, and racial exclusion. In the final contribution of part 1, media scholar June Yeoreum Kim turns our attention to two smart city projects in South Korea and Vietnam. Unearthing these smart city projects, Kim likewise uncovers a long developmentalist, postcolonial history of land clearing as essential for founding smartness. This contribution beautifully shows that smartification processes are both virtual and material: computer ideas and practices of deletion, formatting, and uploading are rebooting processes that are deeply inscribed into South Korean and Vietnamese land.

REFERENCE

Jacobs, Karrie. 2022. "Toronto Wants to Kill the Smart City Forever." *MIT Technology Review*, June 29. www.technologyreview.com/2022/06/29/1054005/toronto-kill-the-smart-city/.

3 The Smartification of Global Development

KENDRA KINTZI AND HILARY FAXON

Smart infrastructures, from networked applications to the increasingly ubiquitous "Internet of Things," dominate our economies and shape our politics. Across the Global South, the past decade ushered in a wave of development projects promoting smart electricity grids, smart transport hubs, climate-smart agriculture, and a variety of interventions that layer new digital technologies into existing rural-urban infrastructures. Powerful donors and multilateral actors repeatedly declare that new technologies are the answer to persistent problems of sustainable development. At the forefront of this wave of smart development, the World Bank promises that a "smart" approach can deliver a "triple win" of maximizing yields, enhancing resilience, and mitigating climate change in agriculture[1] while "integrating urban infrastructure and service delivery to provide solutions to achieve a citizen-centric approach" in smart cities.[2] While the project sites and types of interventions multiply, the unifying promise is clear: smartification is the key to enabling continued economic growth amid rising social, environmental, and climate crises.

This chapter begins with the provocation that smartification has been, since inception, both a political project and a global phenomenon entangled with the postcolonial politics of economic integration and transnational interconnection. Smartification saturates landscapes and troubles analytical categories, transecting the urban and the rural, the agricultural and industrial, the social and the material, and the Global North and South. We situate the emergence of smartification within the longer history of uneven global development, drawing upon insights from critical development studies to understand how smart

---

1 www.worldbank.org/en/topic/climate-smart-agriculture.
2 www.worldbank.org/en/programs/global-smart-city-partnership-program.

infrastructures and projects of smart subject-making take shape within pre-existing social relations of inequality and previous phases of colonial and capitalist expansion. Contrary to the promises above, we argue that smartification's promises of connectivity materialize in a world that is already violently and unevenly interconnected.

Drawing on our ethnographic research in Jordan and Myanmar, we illustrate how the circuitry of smartification intersects with older flows of capital and knowledge. This chapter builds from our efforts to highlight the parallels between global projects of smartification, from smart electricity grid development in Jordan to the building of cell towers on contested land in Myanmar (Faxon and Kintzi 2022). In recent years, technological advances in two-way communication infrastructure have unleashed new waves of investment in smart development, which couple the roll-out of data collection and algorithmic governance technologies with a suite of policy and regulatory reforms to facilitate private investment and the commodification of data. Over the course of our extended field research in Myanmar and Jordan, we witnessed the unfurling of sweeping reforms that dramatically altered the infrastructural landscape, creating new data and investment streams and changing the ways that individuals communicate with each other and the state. In Myanmar, the liberalization of the post and telecommunications monopoly sparked US$2.8 billion in foreign direct investment and a twenty-fold spike in internet users, while in Jordan, renewable energy legislation ushered in over US$4 billion in private investment to support the transition to renewable sources and the digitalization of the country's electricity distribution networks. By analysing these parallel projects of digital-material transformation against the backdrop of global smartification efforts, we highlight both the transnational structures and the situated histories that powerfully shape how projects of smartification come to matter in the world.

The Long Roots of Smart Development

Development agencies, donors, and practitioners increasingly append "smart" to familiar interventions as a way of indexing a new, savvy, efficient approach. In such circles, the mention of "smartification" conjures images of high-visibility projects like Sidewalk Labs' Quayside in Toronto, which purports to "provide a global model for inclusive urban growth,"[3]

---

3 See www.sidewalklabs.com/toronto.

or NEOM in Saudi Arabia, which bills itself as the world's "first cognitive city, where world-class technology is fuelled with data and intelligence to interact seamlessly with its population."[4] Appended as a prefix to a multitude of initiatives and projects, the term "smart" gestures towards the transformative role of digital technologies in solving persistent social, political, and environmental challenges. The term's popularity is apparent in its increasing circulation in the conferences, calls, reports, and white papers that constitute the international development industry. For example, a search of World Bank projects using the term "smart" in June 2023 yielded 331 projects, steadily increasing from the early 2010s. These projects were diverse, ranging from smart villages in Niger to smart cities in Kazakhstan, smart grids in Ukraine and Indonesia to smart governance projects in Mongolia.[5] This very heterogeneity points to the overarching, often unfulfilled, promise of smartness: to solve intractable challenges in disparate places through digital technologies.

Like the word "development" itself, the term "smart" has become an industry buzzword, used in ways that assume a common normative belief, in this case in the power of technology to uplift the lives of the global poor and fix the problems of so-called emerging markets (Rist 2007). This understanding of smartness parallels what Halpern and Mitchell describe as an "epistemology that relies on new practices, technologies, and subjects" (2022, xi). While the promise and perils of digital technologies as tools of growth and governance have recently become a central global concern, the idea that connectivity creates a platform for progress is not new. From Rostow's *Stages of Economic Growth* to Toeffler's techno-optimistic prognosis of humanity's capacity to invent itself out of resource constraints (Koch 2021), the history of post-war global development is rife with models that collapse modernization into a project of linear technological advancement. Today's projects of global smartification – while varied and diverse – often follow in these theoretical footsteps. The goals of reducing inefficiencies, spurring growth, serving citizens, and promoting sustainability are familiar from the past few decades of development practice. What is new is the positioning of networked infrastructures and subjects as the answer to

---

4 See www.neom.com/en-us.
5 See https://projects.worldbank.org/en/projects-operations/projects-list ?os=0&qterm=smart. On June 26, 2023, a total of 22,019 projects were included in this database. While smart appears in a relatively small percentage of these, it is telling that it comprises an increasing proportion: Between June 2022 and June 2023, an additional 103 projects employed the term "smart."

the persistent ills of underdevelopment, financial precarity, and environmental degradation.

Critical development studies (CDS) emerged as a field in the mid-twentieth century in response to the conflation of connectivity with progress, challenging both the assumptions and effects of projects of "development" that fail to recognize the violent and deleterious impacts of colonial and capitalist expansion. As Phil McMichael (2009) has shown, the development project is both an ideology of progress centred around the market episteme *and* a historically specific set of relations that is continually contested, appropriated, and reformulated through the situated struggles of diverse social movements. Today, smartification is an evolution of what Tania Li (2007) describes as rendering technical the process through which development agencies reduce complex social relations, and the political disagreements within them, into clearly bounded objects of improvement. Like preceding development initiatives that depoliticized the conditions that generate underdevelopment, such programs materialize in unexpected ways, with unanticipated politics (Ferguson 1990).

From Dependency Theory (Frank 1966) to World Systems Theory (Cardoso and Faletto 1979; Wallerstein 1974), scholarship in CDS has brought focus to the uneven spatial dynamics of global economic integration, highlighting how the "sites" of development in the postcolonial world are incorporated into power-laden relationships of extraction and exchange. Spatializing contemporary projects of smartification helps us see how uneven relations of accumulation shape the ways that smart infrastructures are constructed, and who benefits.

Contemporary projects of development often employ neocolonial assumptions of ignorance and expertise, in which some people are the agents, and others the objects, of development (Mitchell 2002; Mosse 2004). In contrast, critical scholarship on racial capitalism and feminist political economy (Gibson-Graham 2006; Robinson 2000; Wolford 2021) highlights the pivotal role of subaltern revolutions as key sites of transformative politics and locates the roots of resistance to structures of global exploitation in the culturally rich realms of lived experience (James 1989). From participatory assessments to pluriverse politics (Escobar 1999), critical development scholars are centrally engaged in praxis to enact different futures. Engaging with grassroots mobilizations often brings to light new empirical questions and concerns that challenge existing theories of development (Gago and Mason-Deese 2019) while reflexive and participatory approaches provide pathways to re-imagine and enact smartification in ways that attend to everyday lived experiences in order to cultivate rooted, emancipatory politics.

Redefining Smart Development

Smartification is more globally interconnected and more fragile than is often portrayed. The 2010s were a period in which the promise and perils of digital technologies in relation to development and governance became a central global concern. The resurgence of fiscal austerity and the entrenchment of privatization and liberalization after the 2008 financial crisis simultaneously hollowed out funding for public sector infrastructure around the globe while ushering in waves of private investment and finance capital from a new genre of development actors: the tech "disruptors." Recent ethnographies in China and India have highlighted the key role digital technology plays as a terrain and vehicle for national development and entrepreneurial citizenship (Irani 2019; Lindtner 2020). In Irani's (2019) ethnography of designers and engineers in Delhi, digital connection is essential to what she dubs "rendering entrepreneurial," a move that frames innovation as national improvement and results in individual responsibilization and the subsumption of both hope and radical action.

Below, we draw from separate long-term ethnographic research in Myanmar and Jordan to re-examine smartification from the ground up. These sites provide rich terrain for comparative analysis of how smartification is envisioned, enacted, and inhabited in postcolonial contexts marked by legal pluralism, demographic complexity, and enduring disparities. Both countries have complex histories marked by colonialism, dispossession, and inter-group conflict. Yet both Jordan and Myanmar have been constructed as promising sites of smart development in the past decade. Multilateral projects and public-private partnerships have spurred rapid digitalization and remade agricultural landscapes, energy grids, and everyday livelihoods. In both contexts, euphoric promises of democratization and economic liberalization played key roles in channelling international investments towards smart energy and unbundled telecoms, yet the winners and losers of these processes were shaped by previous waves of imperialism and inequality (Faxon and Kintzi 2022). In Myanmar, digital connection brought millions of people online for the first time and enabled the smart farmer to become imaginable as a development solution, yet rural residents ignored custom-made apps and websites in favour of repurposed social media groups for exchanging agricultural advice or speculating on land (Faxon 2023; Wittekind and Faxon 2023). In Jordan, smart energy investment flows radically remade the country's legal, regulatory, financial, and environmental landscape through new sensors, cables, and monitoring systems designed to cultivate tech-savvy citizens, paid for by private finance (Kintzi 2024).

Bringing our research from these two sites together enables us to discern the broken promises and broader patterns of development's digital turn.

At modest and massive scales, smart development projects index a techno-optimistic orientation that shifts dynamics of power between different actors, favouring hypothetical innovation over acknowledging or addressing structures of racialized and gendered exploitation, extractive accumulation, and dependent economic integration. While multilateral development agencies such as the World Bank play a crucial role in promoting and replicating specific kinds of interventions deemed as "smart," national- and private-level actors are pivotal in defining how smartification materializes on the ground. For example, in an agritech hackathon in Myanmar in 2019, urban youngsters pitched business executives from the nation's largest agrochemical businesses and policymakers on digital solutions in a Swiss-funded impact hub. Notably, no farmers were involved in the multi-day event, which took place in the industrial port area of downtown Yangon. If they had been, the proposals might have been quite different: one social enterprise had recently abandoned agricultural advice apps because farmers preferred to receive SMS reminders. In an interview, one businessman who had served as a judge at the event explained that they did not expect any transformative or profitable solutions to come out of the hackathon but rather participated to raise awareness among youth about agriculture and data science and, more generally, to keep up with digital interventions as a new frontier in their industry. This interaction shows that there can be diverse motivations for the embrace of smartification projects, beyond better or more equitable service delivery. Smart development projects privileged foreign experts, business elites, and entrepreneurial tech "disruptors," even as their proposed solutions sidestepped farmers' demands for access to credit, irrigation, and stable trade.

Similarly, at a solar and smart energy technology exposition in Jordan in the summer of 2022, vendors from around the world showcased their latest innovations in digital metering and electronic monitoring and surveillance technologies. Investors, developers, and utility representatives meandered around the exposition floor, perusing the options for devices and digital platforms that could help automate the grid and improve the efficiency of electricity delivery. The exposition took place inside the air-conditioned convention centre adjacent to the country's flagship business technology park, far from the central and eastern neighbourhoods of the capital city, Amman, where residents often struggle with the skyrocketing cost of electricity bills under the nationwide energy sector reform (Faxon and Kintzi 2022; Kintzi 2024). Since 2018, a growing number of Ammanis have taken to the streets and to social media to protest rising energy prices. In early 2019, protesters ignited the #Not Paying movement, refusing to pay their electricity bills amid rising unemployment and deteriorating economic conditions. Over the course of the COVID-19 pandemic, these conditions worsened demonstrably, particularly for the city's sizeable migrant and refugee populations, who faced rising food insecurity, plummeting household income, and increased difficulty in sustaining household life (UNHCR, UNICEF, and WFP 2020). While the glittering banners of the exposition promised transformative digital technologies, the embodied realities of smart energy transition are experienced in profoundly uneven ways across the city's urban landscape.

Rather than take smart discourses at face value, our work brings attention to the materiality of smartification, highlighting both the entrenchment of old inequalities and the configuration of new forms of extraction. In both Jordan and Myanmar, digital connections materialized within landscapes marked by acute inequality, where histories of colonial and imperial dispossession and contemporary projects of ethnonationalist territorialization had created high degrees of wealth concentration and landlessness. In Myanmar, the very telecom towers that enabled smart farming apps were built in ethnic minority areas marked by long histories of conflict and rural landscapes characterized by decades of military land grabbing. Global projects of smartification rely on rare earth metals and other minerals to create the sensors, catalysts, batteries, and screens (see also Fard, this volume). China dominates the global production of rare earth elements, with over a third of reserves and nearly two-thirds of production in 2020.[6] Escalating demand has fomented a race to find and exploit new extractive sites, whether cobalt mines in the Democratic Republic of Congo or rare earth exploration in Greenland. In Jordan, the embrace of silica mining as a national development strategy is engendering new extractive economies (Madanat et al. 2014) while a recent decision to extend copper mining in a protected nature reserve has generated intense environmental protests under the banner #Save_Dana (Vidal 2021). These geopolitical and geophysical processes introduce new economic opportunities, new forms of environmental contamination, and new social dynamics at the frontiers of smart development.

As each of the examples above illustrates, the smartification of global development is both contested and incomplete. While technology vendors market digital solutions that promise transformative change, these technologies are often packaged together with policy and regulatory

---

6 See www.visualcapitalist.com/rare-earth-elements-where-in-the-world-are-they/.

reforms that accelerate the privatization of infrastructural services and prioritize investor profits. As these technologies are adopted and implemented, they encounter the frictions of lived realities, from Myanmar farmers calling for better irrigation and terms of exchange, to urban residents in Jordan demanding affordable energy and better living conditions. These contestations have the potential to powerfully reshape how smartification unfolds on the ground.

We begin to imagine alternatives by attending to the situated politics of globally interconnected sites. In our own research, we found that smart development projects were adapted, refused, and contested in often surprising ways. In Jordan, the roll-out of smart energy development drove up electricity costs, forcing many residents to switch off their power and turn instead to age-old tactics of thick blankets draped across apartment windows and rugs wedged into the cracks of door frames to keep out the cold. These low-tech, low-cost home remedies are essential tools in sustaining life in the city and call attention to the ways that smart development is intimately bound up with ongoing struggles to find work and survive within the precarious conditions generated by compounding waves of violent dispossession, dependent industrial development, and neoliberal reform. Attending to these situated knowledges and placed-based power geometries is key to understanding the varied, uneven, and deeply contextual effects of globalizing projects of development.

Conclusion

People and places are never just "left out" of development. Rather, connectivity within the circuits of global capitalist exchange and resource extraction generates profoundly uneven effects. Within this context of global interconnection, the imaginaries, infrastructures, and subjects of smartification materialize in specific times, in specific places, and through embodied experiences that shape the possibilities for transformative politics. For rural Myanmar farmers, the construction of a vast cell tower network took place against the backdrop of contested land grabbing and ongoing struggles to maintain agrarian livelihoods. In Jordan, smart energy projects materialize amid ongoing protests for affordable energy and a more equitable distribution of resources and profits. CDS has long critiqued linear conceptualizations of modernization, showing instead how history, geography, and lived ontologies come to matter in shaping how development unfolds – in uneven, contested, and contradictory ways. Bringing these insights into dialogue with critical scholarship on smart systems helps us ground smartification as an evolving project that is taking shape within a world that is already connected by colonial and capitalist relations. Repositioning our vantage point from the "developing" world highlights the ways in which the globe-transforming power of smartification is not a foregone conclusion but rather an unfolding, contradictory, and deeply contested arena of politics.

Grounding smartification in this way presents an opportunity to reflect on the goals of smartification as they are expressed by the powerful private and multilateral actors driving new smart development projects. Rather than delivering on promises of seamless integration, efficiency, and sustainability, smart development imaginaries are built and enacted in postcolonial landscapes marked by profound, enduring colonial inequalities. While these projects do generate new streams of data and revenue, the benefits of smartification accrue asymmetrically. These differentiated impacts are not random but reflect the postcolonial landscapes of particular places and uneven patterns of global development. By situating and spatializing smartification projects within this longer history of uneven development and extractive interconnection, we can both map and potentially rework the power of smartness.

References

Cardoso, Fernando Henrique, and Enzo Faletto. 1979. *Dependency and Development in Latin America*. Translated by Marjory Mattingly Urquidi. University of California Press.

Escobar, Arturo. 1999. "After Nature: Steps to an Antiessentialist Political Ecology." *Current Anthropology* 40 (1): 1–30. https://doi.org/10.1086/515799.

Faxon, Hilary Oliva. 2023. "Small Farmers, Big Tech: Agrarian Commerce and Knowledge on Myanmar Facebook." *Agriculture and Human Values* 40: 897–911. https://doi.org/10.1007/s10460-023-10446-2.

Faxon, Hilary Oliva, and Kendra Kintzi. 2022. "Critical Geographies of Smart Development." *Transactions of the Institute of British Geographers* 47, no. 4 (December): 898–911. https://doi.org/10.1111/tran.12560.

Ferguson, James. 1990. *The Anti-Politics Machine: "Development," Depoliticization, and Bureaucratic Power in Lesotho*. Cambridge University Press.

Frank, Andre Gunder. 1966. "The Development of Underdevelopment." *Monthly Review* 18, no. 4 (September): 17–31. https://doi.org/10.14452/MR-018-04-1966-08_3.

Gago, Veronica, and Liz Mason-Deese. 2019. "Rethinking Situated Knowledge from the Perspective of Argentina's Feminist Strike." *Journal of Latin American Geography* 18 (3): 202–9. https://doi.org/10.1353/lag.2019.0047.

Gibson-Graham, J.K. 2006. *The End of Capitalism (as We Knew It): A Feminist Critique of Political Economy*. University of Minnesota Press.

Halpern, Orit, and Robert Mitchell. 2022. *The Smartness Mandate*. MIT Press. https://doi.org/10.7551/mitpress/14623.001.0001.

Irani, Lilly. 2019. *Chasing Innovation: Making Entrepreneurial Citizens in Modern India*. Princeton University Press. https://doi.org/10.2307/j.ctv941vd8.

James, C.L.R. 1989. *The Black Jacobins: Toussaint L'Ouverture and the San Domingo Revolution*. 2nd rev. ed. Vintage Books.

Kintzi, Kendra. 2024. "The Smart Grid Archipelago: Infrastructures of Networked (Dis)Connectivity in Amman." *Environment and Planning D: Society and Space* 42 (4): 492–511. https://doi.org/10.1177/02637758231209656.

Koch, Natalie. 2021. "The Desert as Laboratory: Science, State-Making, and Empire in the Drylands." *Transactions of the Institute of British Geographers* 46 (2): 495–509. https://doi.org/10.1111/tran.12414.

Li, Tania Murray. 2007. *The Will to Improve: Governmentality, Development, and the Practice of Politics*. Duke University Press. https://doi.org/10.2307/j.ctv11smt9s.

Lindtner, Silvia. 2020. *Prototype Nation: China and the Contested Promise of Innovation*. Princeton University Press. https://doi.org/10.2307/j.ctvz938ps.

Madanat, Marwan, Nidal Mehyar, and Nihaia A. Zurquiah. 2014. *Silica Sand: Mineral Status and Future Opportunity*. Amman, Jordan: Ministry of Energy and Mineral Resources.

McMichael, Philip. 2009. *Contesting Development: Critical Struggles for Social Change*. Taylor & Francis. https://doi.org/10.4324/9780203860922.

Mitchell, Timothy. 2002. *Rule of Experts: Egypt, Techno-Politics, Modernity*. University of California Press. https://www.jstor.org/stable/10.1525/j.ctt1ppnxp.

Mosse, David. 2004. *Cultivating Development: An Ethnography of Aid Policy and Practice*. Pluto Press.

Rist, Gilbert. 2007. "Development as a Buzzword." *Development in Practice* 17 (4–5): 485–91. https://doi.org/10.1080/09614520701469328.

Robinson, Cedric J. 2000. *Black Marxism: The Making of the Black Radical Tradition*. University of North Carolina Press.

UNHCR, UNICEF, and WFP. 2020. "Multi-Sectoral Rapid Needs Assessment: COVID19 – Jordan." https://data.unhcr.org/en/documents/details/75962.

Vidal, Marta. 2021. "Copper Mine Threatens Jordan's Largest Nature Reserve." *Al Jazeera*, October 3. www.aljazeera.com/news/2021/10/3/copper-mine-threatens-jordans-largest-nature-reserve.

Wallerstein, Immanuel. 1974. "The Rise and Future Demise of the World Capitalist System: Concepts for Comparative Analysis." *Comparative Studies in Society and History* 16 (4): 387–415. https://doi.org/10.1017/S0010417500007520.

Wittekind, Courtney T., and Hilary Oliva Faxon. 2023. "Networks of Speculation: Making Land Markets of Myanmar Facebook." *Antipode* 55 (2): 634–55. https://doi.org/10.1111/anti.12896.

Wolford, Wendy. 2021. "The Plantationocene: A Lusotropical Contribution to the Theory." *Annals of the American Association of Geographers* 111 (6): 1622–39. https://doi.org/10.1080/24694452.2020.1850231.

4 Refusing the Urban Laboratory

ABHISHEK VISWANATHAN AND BOBBIE FAN

Introduction

The city called Pittsburgh, Pennsylvania, USA, sits on the lands that have been stewarded by many Native communities, notably the Adena culture, Hopewell culture, Monongahela peoples, Lenni Lenape, Shawnee, and people of the six Nations of the Haudenosaunee Confederacy, namely the Mohawk, Oneida, Onondaga, Seneca, Cayuga, and Tuscarora. In the early decades of the twentieth century, it was a site of Seneca displacement (Diaz-Gonzalez 2020) and industrial expansion, with the 1920s seeing extreme monopolistic wealth accumulation through the manufacturing of steel, glass, and aluminum.

As steel manufacturing began to decline in the 1940s, the city of Pittsburgh began to bet its fortunes on wooing a new "creative class" of scientists and researchers. Richard Florida, a Carnegie Mellon University researcher living in Pittsburgh, authored *The Rise of the Creative Class*, painting Pittsburgh as a "base case" in the transition from an industrial to a "knowledge-based economy" (Florida 2011). This logic drove urban planning that sought to attract this highly mobile "creative class" of educated white men, who would procure lucrative defence and research grants from the US federal government. The wealth from those grants would, in turn, trickle into the surrounding community (Florida 2011). Pittsburgh now brands itself as a city of "Eds and Meds" (Kim 2021), and institutions like The University of Pittsburgh (Pitt) and Carnegie Mellon University (CMU) are spoken of as Pittsburgh's "research and development arm" (Florida 2011). Pittsburgh's turn to this knowledge-based economy, embrace of its universities, and explicit view of the city by its administrators as an "urban laboratory" (Florida 2018) tells a story of academic smartification processes, of academic production of smartness as a tool of speculative land transformation and public space extraction.

These institutions continue the charge of attracting scholars and other creative classes to the city, further expanding into communities at a cost to its current residents, who are erased, displaced, and criminalized (Vitale 2016).

We see the university's role in smartness starkly revealed through two local cases – the "Mon-Oakland" autonomous shuttle and "Crimescan," a predictive policing academic collaboration. Both cases help build an understanding of the mechanisms through which academic production of "smartness" can give epistemological legitimacy to colonial and racist creations of value in property and space that drive urban development. In understanding "smartness" as a vehicle for determining the value of space, these cases also show the crucial need for academic insiders to elevate community conceptions of value as new constructions of "smartness."

Analysis and Approach

For this volume on smartness and processes of smartification, we trace the university's involvement in the production of smartness through the lenses of ongoing settler-colonialism, racial-capitalism, and neoliberalism. We employ a decolonial approach in reframing "smart cities" as a neoliberal incarnation of western municipal authority as founded upon the power to regulate and facilitate commerce (Arnade et al. 2002). We tie the epistemic justification of this authority as creating an idea of the common good that is synonymous with protecting the market to the exclusion and criminalization of "undesirables" (Arnade et al. 2002). We also see the colonial role of academia's production of such epistemic justifications in the self-reinforcing of status quo neoliberalism, with the "political becom[ing] epistemological when any political alternative to the current state of affairs is credibly framed in the same way as fancy against fact or as falsehood against truth" (Santos 2018, viii). This process of knowledge generation is characteristic of how the coloniality of knowledge persists and "flourishes in unexpected and not evident spheres of modern disciplines and academic divisions, in the production and distribution of knowledge, as well as in geo-historical and geopolitical situations" (Tlostanova 2018).

These frameworks provide the basis for our arguments about the techno-solutionism that permeates the university campus and beyond, and the carceral technologies that are born out of that thinking. However, we also recognize that the university does not act alone. The intertwined relationships of the academy, government, non-profits, and corporations work to advance the surveillance state through an opaque system of grants, internal networks, and convoluted justifications couched in academic jargon that can bypass regulatory bodies.

Taking the framing of *The Undercommons* (Harney and Moten 2013) and *A Third University Is Possible* (Paperson 2017), we seek to turn the tools, access, and privilege of the university against institutions and towards uplifting radical and grassroots movement work. At the same time, through contributing our understanding of smart city development to volumes such as this one, we seek to undermine the logic perpetuating techno-solutionism, settlement, and carcerality within the university and within ourselves.

> Universities are land-grabbing, land-transmogrifying, land-capitalising machines. Universities are giant machines attached to other machines: war machines, media machines, governmental and nongovernmental policy machines. (Paperson 2017)

Both of us find our positions at two of the major universities in Pittsburgh to be both precarious and complicit in the extractive "land-transmogrifying" nature of the academy. Though a large ecosystem of universities exists in Pittsburgh, Pitt and CMU have had historic origins as products of industry – via Mellon financial interests and steel baron Andrew Carnegie (Crowley 2005). Both universities boast expansive science and technology departments, which apply for and receive large grants through governmental bodies as well as corporations in a funding ecosystem that prioritizes commercialization and development.

Smartness as an Engine of Public Space Extraction – the Mon-Oakland Connector

In 2019, CMU opened an advanced robotics manufacturing space in a former steel mill turned tech hub called Mill19. The site was developed by Almono Partners, created by the joint investments of four local foundations, the Richard King Mellon (RKM) Foundation, the Heinz Endowments, and two others (Blackley 2019). During the site's ribbon cutting, CMU's president smiled down at a yellow robot arm, symbolizing a city marked by downturn in manufacturing and scrambling for revitalization and rebranding. The irony of turning to autonomous manufacturing in a region devastated by the loss of manufacturing jobs seems to be a footnote to the narrative of tech progress and highlights the forms of labour prioritized in the city's "revitalization."

Emerging alongside the high-tech manufacturing site was a proposal for an autonomous shuttle corridor connecting the two universities

(CMU and Pitt) with the ALMONO development site. Dubbed "The Mon-Oakland Connector," its inception in 2016 came as a part of Pittsburgh's finalist bid for the US Department of Transportation's (USDOT) 2016 Smart City Challenge, which laid out a high-tech vision of the city as shaped by a newly formed conglomerate of government, universities, developers, and tech companies called Smart PGH (Smart PGH 2017). The centrepiece in their proposal was an autonomous shuttle route from the university-centred neighbourhood of Oakland to the new ALMONO high-tech autonomous manufacturing facility (Smart PGH 2017). When the Pittsburgh Smart City Challenge bid lost to other cities, the Urban Redevelopment Authority helped secure state-level Department of Transportation funding to continue the project, despite fierce resident opposition.

The proposed corridor cut through two neighbourhoods, whose residents, since the project's inception, have raised concerns about millions in public funding going to a high-tech shuttle primarily benefiting the university partners (CMU and Pitt) of the ALMONO tech site (TransitCenter 2022).

Pittsburgh's smart city visions have consistently elicited concern from its residents. The development of Google campuses in the East Liberty neighbourhood accompanied a wave of gentrification and mass displacement of Black residents (Davis 2018), while street-level vitriol accompanied the mayor's red carpet welcome of Uber's self-driving cars on Pittsburgh roads (Perrone 2022). However, at public meetings for the Mon-Oakland autonomous shuttle, leaders from the CMU's Metro21 Smart Cities Institute referred to neighbourhood residents fighting the shuttle as backward, lacking understanding of technology, and antiprogress. This condescending characterization of residents showcases an elitism that furthers a colonial vision of progress and conflates high-tech development with smartness and progress (Rodríguez-Alegría 2008). Residents meanwhile highlighted the acute neglect of the region's infrastructure, with cracked sidewalks, food deserts, underfunded transit, and the threat of displacement (Pittsburghers for Public Transit 2019). Residents, along with advocacy group *Pittsburghers for Public Transit*, put forth a counter-proposal called "Our Money, Our Solutions," with concrete proposals for weekend service, grocery store access, and infrastructure repair that residents had been consistently asking for (Pittsburghers for Public Transit 2019).

Meanwhile, ALMONO Partners locally sought to brand Mill19 as an "eco-friendly tech hub." In 2015, alongside RKM Foundation's $4 million for redevelopment, a separate $600,000 grant was awarded under education (Haptas 2016; Richard King Mellon 2015). This marked the creation of CMU's Metro21 Smart Cities Institute, and a conscious coupling of city development with university research production.

Smartness Legitimizing the Violence of Extraction – Predictive Policing

While a portion of Metro21's initial grant went towards the energy infrastructure at Mill19, the rest went toward a "two-year support to conduct projects that use urban predictive analytics to improve quality of life and reduce crime in Pittsburgh" (Richard King Mellon 2015). This was the Pittsburgh predictive policing project, or CrimeScan, our second local case exemplifying smartness as employing carceral logic and tools of smartness in shaping urban space. CrimeScan was deployed in each police district and halved areas for predictive policing for "trial" and "control" (Fitzpatrick et al. 2020). In August 2016, after being developed by researchers at Carnegie Mellon working with the Pittsburgh police chief, predictive policing was piloted in Homewood, a predominantly African American neighbourhood in Pittsburgh. It launched quietly citywide in May 2017.

At a local level, data-driven policing serves the interests of land speculation, by "cleaning up crime," driving up property value, and transforming land for development – a phenomenon well captured by StopLAPDSpying's street-level organizing in Skid Row and analysis of the Los Angeles Police Department's (LAPD) hotspot placements (Stop LAPD Spying 2021). Rather than gentrification leading to displacement, police-driven displacement and violence paves the way for gentrification (Beck 2020).

In CMU Metro21's predictive policing grant and work plan, the researchers under CMU professor Daniel B. Neill touted their predictive services as eliminating "rodent infestations" as well as "emerging hotspots of violence" (Neill and Gorr 2016). This form of hotspot-based policing, targeting locations, with a merged understanding of "cleaning rodents" and fighting crime, was a part of how property values shot up overnight in the East Liberty neighbourhood of Pittsburgh (Numeritics 2015). Here, mass displacements of Black public housing residents occurred simultaneously with the entryway of upscale stores like Whole Foods and Target (Fullilove et al. 2009). Most tellingly, however, Neill's team openly referenced broken windows as a basis for their work in their grant proposal to the RKM Foundation:

> It is well established that signals of urban disorder (e.g., "broken windows") can lead to or attract criminal behavior that hardens over time. Therefore

the two areas proposed for further funding are intertwined; for example, 311 calls for abandoned, run-down buildings and properties are not only targets for clean-up but also likely leading indicators of crime. (Neill and Gorr 2016)

This presents a clear connection between criminalization and clean-up, with the goal of shaping public space for private ventures via policing. Furthermore, the project uses the "smart" vehicle of machine learning, which takes past location-based indicators as future signals. This place-based logic of policing flows from broken windows theory, which legal scholar Bernard E. Harcourt traced from a popularized 1982 article based on Philip Zimbardo's 1969 flawed (and staged) experiment on vandalism (Zimbardo 1972; Harcourt 2001). Broken windows then underpinned "Stop and Frisk" policies that failed to account for any shifts in overall crime (Harcourt 2001) but where 87 per cent of searches landed on Black and Latino populations (Chavis 2014). This illustrates a circular relationship, where academic scholarship erroneously linking disorder and crime led to policies that ultimately criminalized disorder (Harcourt 2001) and expanded inequities in the carceral system.

Predictive policing and gang databases have found their way into almost every major metropolitan area in the US, with data-driven policing rolled out as both a policing operations boost, as well as a move towards a world of reformed policing (McQuade 2019). The pervasive datafication of public and private services has enabled a new surveillance infrastructure allowing for greater control of the economically disenfranchised via a pervasive collection of data to flag and identify locations, communities, and other populations (McQuade 2019).

CMU Professor Daniel B. Neill had previously deployed his broken-windows-based CrimeScan with the Chicago Police Department in 2009 (Richardson et al. 2019). PredPol, an early predictive policing vendor, was funded by university military research (Stop LAPD Spying 2021). The proliferation and export of such products at a municipal level shows a replication of academic complicity in both legitimizing the logic of policing and providing ever-evolving "smartness" value to imbue in policing products (Stop LAPD Spying 2023). Universities benefit from this relationship in the commercialization of their research. Lab-grown software, in addition to its outsourced grant-funded price tag, also allowed skirting around pesky procurement processes – both huge benefits to government officials and openly boasted about by Pittsburgh's then-mayor in reference to the city of Pittsburgh's data-sharing memorandum of understanding with CMU: "I don't have to put out an RFP [Request for Proposals]. I can just pick up the phone, call the university, and say, 'I need your team to develop this for me'" (Florida 2018).

Academics give an objective veneer to new brands of policing by establishing rationally proven, journal-backed theories, policies, and technology products. In doing so, they erase the history that populates the crime data powering predictive policing models. This history extends from slavery era laws and practices to redlining and the War on Drugs campaign, which resulted in mass impoverishment and criminalization of Black communities (Taifa 2021).

CMU presented CrimeScan as an early form of "Tech for Good" work (Neill 2013), with a need to both pilot the work and experimentally "prove" the efficacy of the "treatment" of hotspots with additional police patrols. The selection of its pilot location was determined in secret – the predominantly Black neighbourhood of Homewood (Neill 2016). Community representatives were never consulted, and the pilot launched in 2017, with a citywide launch a few months later. The "treatment" program involved slicing each police zone geographically in half, with additional police patrols only sent to "treatment" zones. At the end of a year period, they compared the crime statistics between these zones to examine the efficacy of their predictive policing algorithm (Fitzpatrick et al. 2020). Not only did the research team and the city approach experimentation with secrecy, but they also chose a Black neighbourhood as a testing site for predictive policing, evoking the historic use of Black populations as sites of academic and medical experimentation.

Rejecting Academia's Reformism

The primary aim of our work is to show the mechanisms that universities can and do employ in legitimizing smartness as a framework for public space extraction. Our secondary aim is to highlight the fugitive role of the insider within academia, to continue the work of the Third University. In the framing of cities as urban laboratories, universities devalue the lived experience, knowledge, and collective power of communities, which we view as a source of relational knowledge or "smartness" resisting top-down formulations of the term.

As organizers within academia, we caution against trusting and following reformist institutional agendas, recognizing the ways they continue to transform productions of "smartness" to serve colonial or carceral aims. In Pittsburgh, this has involved smartification processes and the use of "smartness" to push a gentrifying, high-tech autonomous shuttle while ignoring community needs or to justify and concentrate policing in overpoliced places.

We instead invite readers to a shared work of accountability by placing researchers in accountable relationships with place and people, allowing for opportunities to turn tools against institutions and towards uplifting radical movement work.

Community resistance in our first case study on the Mon-Oakland Connector was led by *Pittsburghers for Public Transit* (PPT) as well as a coalition of residents who sought campus member allyship to counteract university support for the project as well as against characterizations of residents as ignorant of technology. The city responded to initial resident concerns around autonomous technology by soliciting a Knight Foundation grant to "demystify" self-driving cars (Murray 2019). In response to the perception that resident concerns could be addressed by "demystification," organizers rallied researchers to testify at the hearing on the grant (Deto 2019) and solicited a literature review on safety and labour impacts of autonomous tech with University of Pittsburgh researchers (Amruthapuri and Bartel 2019), which eventually led the city to strike autonomous shuttles from its proposal. A second report, The *People's Audit of the Mon-Oakland Connector*, was produced by students at *Tech4Society*, a student-led organization at CMU supporting grassroots and advocacy organizations. The report audited the Mon-Oakland Connector's proposal, finding the local universities (CMU and Pitt) to be the primary beneficiaries of the shuttle, while laying out more effective transit alternatives sourced from the collective visions put forward in the resident proposal "Our Money, Our Solutions" (Fan et al. 2020). The report helped build media momentum against the shuttle (Yablonsky 2020), and enough popular outcry gave political momentum to propose transferring funds from the project to community benefit proposals (TransitCenter 2022). The shuttle proposal became a key mayoral campaign issue, and the next elected mayor finally halted and transferred funds from the project (Conway 2021).

Resident-researcher solidarity was also key in organizing against predictive policing. The *Coalition Against Predictive Policing* (CAPP-PGH). CAPP-PGH, emerged from *Tech4Society* after responding to concerns brought forth by a local privacy advocacy group about technology developed at the same university they attended (CMU). The group focuses on solidarity building with Black activists in the region and counter-researching and exposing the harms and secrecy of the CMU Metro21 predictive policing project. After Pittsburgh's Black Activist Coalition named ending predictive policing in their twelve demands on policing during the George Floyd uprisings, the city of Pittsburgh ceased using predictive policing in June 2020 (Lord 2020). While pushing for accountability within the university by gathering over three thousand signatures with demands for the university to "Confront Racist Policing in Our Community" (Coalition Against Predictive Policing 2020), CAPP-PGH also organized legislatively to bypass university authority, leading to a bill regulating the use of facial recognition and predictive policing (Rihl 2021).

Universities have long argued that they are a force for good within the ecology of civic space and that they deserve unquestioned public support to continue to carry out their missions. In cities like Pittsburgh, whose leaders have long subscribed to the "Creative City" script, the neoliberal positioning of the university as a cure-all for social ills, as well as the carefully constructed image of the university as a place of constant progress, allows the university to operate almost unilaterally in deploying research projects and designing curricula. By developing a close-knit relationship with government through project partnerships, providing seats on the board of trustees, and crafting funding mechanisms, the two institutions are able to mutually reinforce each other's positions as institutions for good. This quid pro quo arrangement is steeped in bureaucratic legalese and academic jargon but also presents a public-facing side that insists that only through a consortium (of universities, government bodies, corporations, and non-profits) can city issues be addressed.

We invite readers to Kaba's abolitionist practice of refusal, care, and collectivity.

> Refusal: because we cannot collaborate with the prison-industrial complex, as "only evil will collaborate with evil" (June Jordan). Care: because "care is the antidote to violence" (Saidiya Hartman). Collectivity: because "everything worthwhile is done with others" (Moussa Kaba). (Kaba 2021, 23–4)

References

Amruthapuri, Rahul, and Sinjon Bartel. 2019. "Pittsburghers for Public Transit." July 10. www.pittsburghforpublictransit.org/wp-content/uploads/2019/07/PPT-AV-paper.pdf.

Arnade, Peter, Martha Howell, and Walter Simons. 2002. "Fertile Spaces: The Productivity of Urban Space in Northern Europe." *The Journal of Interdisciplinary History* 32 (4): 515–48. https://doi.org/10.1162/002219502317345493.

Beck, Brenden. 2020. "The Role of Police in Gentrification – The Appeal." *The Appeal*, August 4. https://theappeal.org/the-role-of-police-igentrification-breonna-taylor/.

Blackley, Katie. 2019. "Mill 19 Is a High-Tech Facility with a History of Major Manufacturing." *90.5 WESA*, September 5. sec. Development & Transportation. www.wesa.fm/development-transportation/2019-09-05/mill-19-is-a-high-tech-facility-with-a-history-of-major-manufacturing.

Chavis, Kami. 2014. "The Legacy of Stop and Frisk: Addressing the Vestiges of a Violent Police Culture." *Wake Forest Law Review* 49 (January). https://papers.ssrn.com/sol3/papers.cfm?abstract_id=2873807.

Coalition Against Predictive Policing. 2020. "CMU: Confront Racist Policing in Our Community." *Action Network*, June 10. https://actionnetwork.org/petitions/cmu-confront-racist-policing-in-our-community.

Conway, Brian. 2021. "Pittsburgh Mayor-Elect Ed Gainey to Place Controversial Mon-Oakland Connector Project under Review." *Technical.Ly*, November 24. https://technical.ly/civic-news/mon-oakland-connector/.

Crowley, Gregory J. 2005. *The Politics of Place: Contentious Urban Redevelopment in Pittsburgh*. University of Pittsburgh Press. https://doi.org/10.2307/j.ctt7zw8zt.

Davis, Jeremiah. 2018. "What's Left When the Gentrifiers Come Marching in." *Publicsource*. www.publicsource.org/whats-left-when-the-gentrifiers-come-marching-in/.

Deto, Ryan. 2019. "Here's What Experts Say Pittsburgh Should Implement Instead of Driverless Cars." Pittsburgh City Paper, July 26. www.pghcitypaper.com/news/heres-what-experts-say-pittsburgh-should-implement-instead-of-driverless-cars-15494658.

Diaz-Gonzalez, Maria. 2020. "The Complicated History of the Kinzua Dam and How It Changed Life for the Seneca People." *The Allegheny Front*, February 12. www.alleghenyfront.org/the-complicated-history-of-the-kinzua-dam-and-how-it-changed-life-for-the-seneca-people/.

Fan, Bonnie, Dave Ankin, Sarah Kontos, and Satvika Neti. 2020. "Pittsburghers for Public Transit." April 10. https://www.Pittsburghforpublictransit.Org/Wp-Content/Uploads/2020/04/PPT-Mon-Oakland-Final-Report.Pdf.

Fitzpatrick, Dylan J., Wilpen L. Gorr, and Daniel B. Neill. 2020. "Policing Chronic and Temporary Hot Spots of Violent Crime: A Controlled Field Experiment." *arXiv*. https://doi.org/10.48550/arXiv.2011.06019.

Florida, Richard L. 2011. *The Rise of the Creative Class*. Basic Books. https://doi.org/10.4337/9780857936394.00008.

Florida, Richard L. 2018. "Pittsburgh's Bill Peduto on AVs and an Economy for All – Bloomberg." *CityLab*, February 8. www.bloomberg.com/news/articles/2018-02-08/pittsburgh-s-bill-peduto-on-avs-and-an-economy-for-all.

Haptas, Maya. 2016. "Almono's Mill 19 Will Transform into Eco-Friendly Tech Hub While Preserving Steel Skeleton." *Nextpittsburgh*. https://nextpittsburgh.com/city-design/mill-19-hazelwood/.

Harcourt, Bernard E. 2001. *Illusion of Order: The False Promise of Broken Windows Policing.* Harvard University Press. https://doi.org/10.2307/j.ctv21ptzhm.

Harney, Stefano, and Fred Moten. 2013. *The Undercommons: Fugitive Planning & Black Study.* Minor Compositions. https://doi.org/10.5070/H372053213.

Kaba, Mariame. 2021. *We Do This 'Til We Free Us: Abolitionist Organizing and Transforming Justice.* Vol. 1. Haymarket Books. https://doi.org/10.1007/s12111-022-09605-2.

Kim, Joshua. 2021. "'The Next Shift': From Manufacturing to Meds (and Eds)." *Inside Higher Education,* August 19. www.insidehighered.com/blogs/learning-innovation/%E2%80%98-next-shift%E2%80%99-manufacturing-meds-and-eds.

Lord, Rich. 2020. "Black Activist Collective Demands Pittsburgh and County Commit to a Dozen Changes on Policing within a Week." *PublicSource,* June 15. www.publicsource.org/black-activist-collective-demands-pittsburgh-and-county-commit-to-a-dozen-changes-on-policing-within-a-week/.

McQuade, Brandon. 2019. *Pacifying the Homeland.* University of California Press. https://doi.org/10.1525/9780520971349.

Murray, Ashley. 2019. "Grant to 'Demystify' Self-Driving Vehicles Draws Skepticism." *Pittsburgh Post Gazette,* May 6. www.post-gazette.com/local/city/2019/05/06/Knight-Foundation-grant-to-help-Pittsburgh-teach-residents-about-self-driving-car-technology/stories/201905060086.

Neill, Daniel B. 2013. *Machine Learning and Event Detection for the Public Good,* 36. Carnegie Mellon University. https://www.nsf.gov/awardsearch/showAward?AWD_ID=0953330.

Neill, Daniel B., and Wil Gorr. 2016. "Work Plan for Mellon Grant." *City of Pittsburgh.* https://capp-pgh.com/files/RTK_Request.zip.

Numeritics. 2015. "White Paper: East Liberty Revitalization: Crime Strategy Implementation." East Liberty Development, Incorporated (ELDI). https://eastliberty.org/wp-content/uploads/2016/12/White-Paper-3-Crime-Strategy-Implementation.pdf.

Paperson, la. 2017. *A Third University Is Possible.* University of Minnesota Press. https://doi.org/10.5749/9781452958460.

Perrone, Alex. 2022. "Pittsburgh Is Falling Out of Love with Uber's Self-Driving Cars." *Endurance Warranty.* www.endurancewarranty.com/learning-center/tech/pittsburgh-hates-uber-self-driving-cars/.

Pittsburghers for Public Transit. 2019. "Sign the Petition: Our Money. Our Solutions." https://actionnetwork.org/petitions/our-money-our-solutions.

Richard King Mellon. 2015. "2015 Annual Report." Richard King Mellon Foundation. https://archive.org/details/rkmf-annual-report-2015.

Richardson, Rashida, Jason M. Schultz, and Kate Crawford. 2019. "Dirty Data, Bad Predictions: How Civil Rights Violations Impact Police Data, Predictive

Policing Systems, and Justice." *New York University Law Review* 94 (April): 15–55.

Rihl, Juliette. 2021. "Pittsburgh Council Votes to Regulate Facial Recognition, Predictive Policing." *PublicSource*, February 12. www.publicsource.org/pittsburgh-city-council-vote-regulate-facial-recognition.

Rodríguez-Alegría, Enrique. 2008. "Narratives of Conquest, Colonialism, and Cutting-Edge Technology." *American Anthropologist* 110 (1): 33–43. https://doi.org/10.1111/j.1548-1433.2008.00006.x.

Santos, Boaventura de Sousa. 2018. *The End of the Cognitive Empire: The Coming of Age of Epistemologies of the South.* Duke University Press. https://doi.org/10.1215/9781478002000.

Smart PGH. 2017. "Once More into the Future Dear Friend: City of Pittsburgh for the US DOT Smart City Challenge." U.S. Department of Transportation. www.transportation.gov/sites/dot.gov/files/docs/Pittsburgh-SCC-Technical-Application.pdf.

Stop LAPD Spying. 2021. "Automating Banishment." https://automatingbanishment.org/.

Stop LAPD Spying. 2023. "Academic Complicity." https://stoplapdspying.org/academic-complicity/.

Taifa, Nkechi. 2021. "Race, Mass Incarceration, and the Disastrous War on Drugs." Brennan Center for Justice, May 10. www.brennancenter.org/our-work/analysis-opinion/race-mass-incarceration-and-disastrous-war-drugs.

Thompson Fullilove, Mindy, Lourdes Hernádez-Cordero, and Robert E. Fullilove. 2009. "The Ghetto Game: Apartheid and the Developer's Imperative in Postindustrial American Cities." In *The Integration Debate*, edited by Chester Hartman and Gregory Squires. Routledge. https://doi.org/10.4324/9780203890462-19.

Tlostanova, Madina. 2018. "What Is Coloniality of Knowledge?" Bloomsbury Visual Arts. http://urn.kb.se/resolve?urn=urn:nbn:se:liu:diva-153604.

TransitCenter. 2022. "Pittsburghers Versus the Machine – TransitCenter." https://transitcenter.org/pittsburghers-versus-the-machine/.

Vitale, Patrick. 2016. "Cradle of the Creative Class: Reinventing the Figure of the Scientist in Cold War Pittsburgh." *Annals of the American Association of Geographers* 106 (6): 1378–96. https://doi.org/10.1080/24694452.2016.1199317.

Yablonsky, Dan. 2020. "News Roundup: New PPT Report Gets to the Heart of the Mon-Oakland Connector." *Pittsburghers for Public Transit*, April 21. www.pittsburghforpublictransit.org/news-roundup-new-ppt-report-gets-to-the-heart-of-the-mon-oakland-connector/.

Zimbardo, Philip G. 1972. *A Social-Psychological Analysis of Vandalism: Making Sense of Senseless Violence.* National Technical Information Service.

5 Making Timber "Smart": Architecture, Technology, and Place in the Pacific Northwest

MEG WIESSNER

Artist Statement

In response to a growing alarm about the climactic impacts of concrete and steel, architects, corporations, and governments are investing in engineered wood products as alternative, low-carbon structural materials. Imagined as both ecologically and technologically advanced, mass timber architecture evokes multispecies forest landscapes even as its uptake depends on developments in building information modelling and digital manufacturing. Smartness often implies that what came before was not smart – backward, imperfect, compromised – but as others in this volume point out (Carlan; Kintzi and Faxon), smartness can also absorb and metabolize what came before, making old relations appear new again. My research explores the continuity of these smartification processes by showing how the historically extractive timber economy of the Pacific Northwest of North America is being reconstructed as a "smart" industry thanks to interest in low-carbon architecture. Against associations with ecological degradation and rural decline, timber is repositioned as part of a new, technologically sophisticated circular economy. This reimagination rests on a set of intersecting but contested claims: that digital prefabrication of wood panels off-site will allow for more efficient construction, accelerating the productization of the built environment; that this will bring advanced manufacturing to rural communities while solving the urban affordable housing crisis; and that this "renewable" material can sequester carbon at scale in buildings and landscapes. Smartification, here, extends beyond datafication, tracking, and monitoring (even as it depends on such processes) to encompass wider ideas about resilience, ecological management, and the warming climate as a systems problem rather than as a political or economic reckoning. Even those who challenge the timber industry's claims that wood use is environmentally benign feel compelled to engage on the same semiotic terrain, calling for wood for these projects to be sourced from climate-smart forestry.

These photographs were taken during fieldwork in Oregon and Washington in 2022. Alongside writing, I use photography as a communicative register for conveying the stakes and scale of this industry. Media production around environmental design sometimes fetishizes techno-natural hybrids, generating affective enthusiasm for the instrumentalization of non-human life (Halpern 2022; Johnson and Goldstein 2015). These images, then, try to avoid fetishizing non-human life even if they depict it. I have also tried to think about what it means to use images critically in research on architecture – a field where visuals constantly risk legitimating rather than unsettling current regimes of production. My hope is that these images of timber's supply chain depart enough from the visual conventions of contemporary design photography to change the frame of reference from which people approach timber as an architectural material.

The rise of mass timber, while fuelled by climate neutrality pledges and technological change, continues to be shaped by patrilineal settler family businesses in logging and milling; by assertions of tribal sovereignty and overdue recognition of Indigenous forest management; by the legacy of forest defence and environmental litigation; and by forests themselves as they burn and change. Talk of digital workflows, precision fabrication, and advanced construction is inseparable from the region's histories of dispossession, logging, fire suppression, labour struggle, and racial exclusion. These particular photos were taken in the traditional homelands of the confederated tribes and communities of the Grande Ronde, Warm Springs, Siletz, Colville, Kalispel, and Yakama Nations.

Figure 5.1. This family-owned sawmill and cross-laminated timber (CLT) manufacturer in Washington – a prominent local employer – specializes in the small-diameter logs used in mass timber products. The company is part of a local forestry collaborative group including industry reps, the USDA Forest Service, Indigenous resource managers, and environmentalists, and it has used thinning contracts on a National Forest to supply the CLT plant.

Figure 5.2. The Peavy Hall Forest Science Complex at Oregon State University is equipped with over 350 sensors as part of the SMART-CLT Project to monitor moisture, structural displacement, weather, and vibration. The architects claim 1,884 metric tons of carbon dioxide have been sequestered in the CLT, glulam, and mass plywood panels, primarily sourced from Oregon forests.

Figure 5.3. A mural in Mill City, Oregon, down the road from an engineered wood plant. The mill towns of the Santiam Canyon were at the heart of some of the most visible confrontations during Oregon's "Timber Wars" and forest defence actions in the 1990s. Several of these towns were devastated during the 2020 Labor Day Wildfires.

Figure 5.4. Cross-laminated timber (CLT) was pioneered in Austria and the Alps in the 1990s. Manufacturing equipment, like this timber press in Oregon, often comes from Germany and Italy. This new facility was opened by a family business specializing in veneer and plywood, and which has been the target of recent environmental direct action related to their post-fire harvesting practices.

Figure 5.5. View of Pahto or Mount Adams from within the lands of the Yakama Nation, with the 2015 Cougar Creek wildfire burn visible across the southeastern slope. The 1855 treaty between the Yakama and the United States stipulated that the latter would provide a sawmill, but they never did. The Yakama opened their own sawmill in 1999, and like other nations in the region, have been approached by firms looking to source mass timber components from tribal enterprises.

Figure 5.6. This robot at a mass timber fabrication shop in Portland is pictured predrilling holes in a glulam beam for screws to be installed on-site during construction. Reducing erection time and on-site labour expenses through prefabrication are major selling points for mass timber projects.

Figure 5.7. Wood for these glulam arches – a centrepiece of Portland's new airport roof – was sourced from the Yakama Nation before being manufactured by a family business in Eugene and fabricated at another shop in Portland.

Figure 5.8. Bookstore display in Portland, Oregon.

References

Halpern, Orit. 2022. "Against Catastrophe." Presented at the Smartification of Everything, University of Ottawa, March 9. www.youtube.com/watch?v=QG782t0fGUE.

Johnson, Elizabeth R., and Jesse Goldstein. 2015. "Biomimetic Futures: Life, Death, and the Enclosure of a More-Than-Human Intellect." *Annals of the Association of American Geographers* 105 (2): 387–96. https://doi.org/10.1080/00045608.2014.985625.

6 Unformattable: The Meaning of Cleared Land in Copy-Pasteable Smart City Technology Transfer from South Korea to Vietnam

JUNE YEOREUM KIM

Here is a scene from a smart city launching event that took place on December 31, 2020, in Ho Chi Minh City, Vietnam: the leaders of the Central Committee of Ho Chi Minh City perform a symbolic city launching ceremony for Thủ Đức City, a new urban zone in Ho Chi Minh City. By pushing the button attached to the round objects – that look like crystal balls with pixelated texts on them – placed in front of each committee member, they completed their launch of the city, which was actually a 5G network (Thế 2020). This event promulgated the establishment of Thủ Đức City as a part of Vietnam's national plan to transition to a digital society; at the same time, this event marked the launch of 5G networks in Vietnam for popular use.

The Thủ Đức City launch captures the popular understanding of "smartness" in the beginning of the 2020s. From the way the event is staged, we can see that cities and infrastructures based on "smartness" are distinctively understood as something that are associated with digital and intangible technologies. In contrast to stereotypical launch events for supposedly non-smart places that take place at physical sites with people shovelling a scoop of earth to symbolize physical construction, launching a smart city in Thủ Đức City is regarded as equivalent to establishing a 5G network. The event demonstrates that the main ingredient for the "smart" city is considered to be the high-speed internet network that is not immediately visible or tangible, instead of dirt and actual land. Behind this performative ceremony lies an assumption that smartness exists purely in digital forms, grounded in the airwaves. Considering the other side of this popular imagery of smartness as largely intangible and immaterial, this essay brings attention to underlying issues such as physical costs, multinational actors, and local-specific aspirations that are easily masked by standardized and technological systems of smartness.

By putting the supposedly universalizing process of digitalization into a transnational context, this essay highlights the contact zone where digitalization meets diverse physical entities and conditions. This discussion shows that the smart city movement cannot simply be considered as a global phenomenon merely consisting of technical systems; it is an entanglement of histories and aspirations grounded in specific, physical places.

Against the popular understanding of smartness as unrooted from physical space represented by the performative ceremony for Thủ Đức City, I argue that erased land is a core element of where smartness is grounded. Smartness is not only grounded in the technologies and infrastructures that run the system but is more fundamentally based on land as geophysical mass. In the case study of the transnational "export" of smart cities from South Korea to Vietnam, I attend to the material precondition of smartness, that is, the physical manipulation and erasure of land. While smart cities are considered to be based on copy-pasteable formulae as noted in the concept of "test-bed urbanism" by Orit Halpern et al. (2013), I build on their analysis to highlight the fact that the practice of "deleting" or clearing pre-existing elements on land precedes copy-pasting, and that the "deleting" in this case is not merely a computational imaginary but is an actual large-scale practice grounded in local-specific histories, material conditions, cultural and economic aspirations, and geophysical manipulation.

Especially when considering the transnational context where the smart city as a template is transferred and implemented in diverse geophysical and sociocultural settings, focusing on the erased land reveals the oft-masked aspirations of those in power who advance the trends of smartification throughout the globe. In this essay, I posit smartness as the operating logic of ubiquitous computing technologies and systems that work towards the goal of hyperconnection and optimization of things that are sensed and processed by the system.

Geophysical Redesign for Smartness

Smartness is not a mere idea that exists in the air. Rather, it's a whole system – assemblages of infrastructure and hardware that involve multinational actors. The ceremony in Thủ Đức City involved numerous hidden actors behind the scenes. The 5G network in Thủ Đức City was powered by Viettel, the largest telecommunications service provider in Vietnam, owned by Vietnam's Ministry of Defence. In October 2020, Viettel joined forces with Vingroup, Vietnam's top conglomerate, to develop 5G technologies. Additionally, Thủ Đức City and some of its smart buildings and

Figure 6.1. A piece of cleared land in Thủ Đức City, where a smart city just began to be built in 2022. Smart cities are idealized, experimental forms of urban environment that work best when they are realized in equally idealized, experimental environments such as newly developed urban zones that are built from scratch on cleared land. Image source: Author.

systems were developed in collaboration with the South Korean government and corporations. The construction of Thủ Đức City closely aligned with the South Korean government's "New Southern Policy" (신남방정책, Shinnambangjŏngch'aek), a diplomatic policy towards ASEAN (Association of Southeast Asian Nations) countries. Announced in 2017, the policy represents South Korea's attempt to take precedence over China and Japan by reinforcing its relationships with Southeast Asian countries. Several government-led initiatives for exporting smart city technologies are aligned with this policy, which provides solutions for urban planning, building information and communications technology infrastructures, and establishing business relations (Ministry of Culture, Sports and Tourism of South Korea 2021). With this complex background that involves many different parties, for the smart technologies to be smoothly transferred and implemented, it was assumed that the South Korean city formula was applicable to other non-Korean cities. But what is needed for the Korean city to be exported and implemented in a non-Korean city context? Among many conditions, I argue that erasing or rebooting the land is one of the essential prerequisites for smartness to be enacted.

It might seem that the land is cleared only for technological purposes. Indeed – in a smart city, physical entities are connected to sensors and then rendered into data that is stored and managed via cloud computing. High speed connection (as of 2022, 4G technology is what allows high speed connection on a popular scale) between sensors – that detect numerous human and natural activities in the surroundings – and the computing system is the core requirement for realizing a smart city. Because existing structures such as buildings, homes, mountains, and islands can interrupt 5G communications, building the environment from scratch can be considered as an ideal way to create smart environments – geophysical environments are redesigned and maintained for seamless 5G communication, ultimately for the optimized operation of smartness. Through this rebooting process, land is cleared and prepared to be detected, processed, and uploaded through smart technologies.

Examined in the context of transnational transfer of smart city technologies, we can observe that clearing land precedes smartness being circulated and implemented in the "air" across national borders. In order for the abstract idea of smartness to be implemented in the air, it is required that physical land is manipulated. Juxtaposing the photographs from the construction sites of Songdo in South Korea and Thủ Đức City in Vietnam shows that the two cities are not easily distinguishable from one another. Both cities are built from scratch, from the vast empty land that is not typically found in already established cities around the world. In order for smart cities to be exported and imported as supposedly context-free and replicable city-making templates based on smartness, they are being built upon erased land.

This land clearing is not a new practice that emerged with the need to install smart technology infrastructures and devices. I grew up in South Korea from the 1990s to the 2000s and have been observing changes in Vietnam throughout the 2010s, so the land clearing in smart city construction sites seemed familiar. The spectacle of cleared land in preparation for the operation of smart technologies in Vietnam evokes chains of historical and cultural memories I have learned about and witnessed directly in the history of postwar industrialization and development in Korea since the 1960s. There seemed to be certain similarities between the cleared land in the blasting of Bamseom Island for the development of Yeouido financial hub in the 1960s in Korea and the cleared land for the development of the smart urban area of Thủ Đức City in the 2020s in Vietnam. More than the visual resemblance and the technical fact that the pre-existing lives and structures on the land were erased, there are shared and hidden aspirations for newer, profitable futures behind these land clearings across time.

The land clearing that followed the idea of transferring smart cities from one country to another is grounded in a deeper historical pattern of development – economic development through the transference and implementation of technological knowledge, to be specific. Given the nature of smart cities being the product of multinational and multicorporate collaboration, it is difficult to pinpoint one exact source of influence for Thủ Đức City. However, it is still true that the technology transfer of smart cities by the South Korean government through the "K-City Network" initiative includes Vietnam as one of the main recipients (Ministry of Land, Infrastructure and Transport of South Korea 2021), and that this technology transfer has existed in different iterations in the past such as KSP (Knowledge Sharing Program) and Saemaul ODA (Official Development Assistance). These technology transfer initiatives are rooted in the South Korean development model that was formed during the rapid industrialization of postwar, authoritarian South Korea during the 1960s and the 1970s (Im and Choi 2017, 182–3, 185–6; Lee 2021, 55), which prioritized economic development.

Local-Specific Meanings of Smartness

Do spaces based on smartness hold the same meaning regardless of their geographic locations? The transnational transfer of a smart city from South Korea to Vietnam allows for an analysis of the discursive aspect that generates distinctive, local-specific versions of smartness. In addition to having its material preconditions masked by its misleading immateriality, the ambiguous rhetoric of "smart" also mispresents the outcome of the implementation, which may not necessarily be promising. While ambiguous, "smart" as in smart technologies and smart cities in South Korea and Vietnam in the early twenty-first century holds specific connotations: the prosperity of the whole nation, which will ultimately benefit its people, and leveraging economic "backwardness" in the digital age where most things, including mainstream economic and governmental systems, can be built on the supposedly intangible world of digital. This is related to why erasing land for smartness is easily condoned in South Korea and Vietnam. While the practice of erasing land for smartness can be situated within a longer genealogy of urban redevelopment, displacement of residents, and eviction involving conflict and violence, new redevelopment projects and cleared land can still be easily spotted in urban areas in South Korea and Vietnam. There exists this consensus of sorts that land clearing is inevitable because it can be justified for the greater good or because the people who have power to decide to clear land or promote smartification will make it happen in the end, no matter what.

Smartness is often misunderstood as a context-free, replicable template. Orit Halpern et al. (2013) observe this phenomenon as "test-bed urbanism." A concept developed around the urban environment and life shaped by an algorithmic environment in Songdo, a smart city in South Korea, "test-bed urbanism" discussed by Halpern et al. refers to the urban environment-making practice where urban units are built "from scratch" to function as a "controlled and isolated development environment in which the operability of new technologies are tested for large systems" (Halpern et al. 2013, 274–5). Developed as a template based on "test-bed urbanism," Songdo has been exported to other parts of the world, including Thủ Đức City in Ho Chi Minh City, Vietnam. The purpose of my analysis is not to show how "test-bed urbanism" is proven to be true in the smart city export from South Korea to Vietnam. Instead, it is to critically engage with the concept's contribution to explaining the proliferation of smart cities based on the supposed sameness of cities in different physical settings. I further build on the concept by applying it in a transnational context in which going digital and platformization of society hold local-specific implications.

While Halpern et al.'s (2013) concept of "test-bed urbanism" is based on the idea that a piece of land can be considered an experimental laboratory, I expand on this conceptualization to regard the "test-bed" or the smart city as a space where computational understanding of the world prevails. Land clearing for building smart cities also evokes the usual practice in computer use of creating new documents from scratch, duplicating them, and deleting them. In a video game, if your gameplay did not go well, you have the option to not save it so that you can replay it. Format devices if you want clean slates. Reboot the machine if it does not work well. Widespread use of digital media made users adopt the idea that you can easily start things over from scratch. Similarly, for smartness, land is regarded as something that can be formatted or rebooted as in computers or video games.

However, the clearing of physical land that is a prerequisite for "test-bed urbanism" is more than a notion based on digital mindset – it requires the actual rearrangement of physical land and people and things on it, which is not an easy task and involves time, conflict, compromise, and sacrifice. Still, "test-bed urbanism" is possible, especially in places outside the Global North. Meanwhile, smartness is chosen because it appears to offer solutions to the urgent developmentalist needs of societies. In the

Figure 6.2. A screenshot of the promotional video of the city-building simulator game "Cities: Skylines." The players can manipulate functions and parameters in the game to experiment with actual issues in urban planning, such as zoning, road placement, planning public services, public transportation, and so on. Smart cities, especially the ones built on cleared land, are based on a similar concept and imagery – urban planning elements can be copied, pasted, tried out, and deleted as if in a video game. In April 2018, South Korea's Land and Housing corporation (LH), with support from the Ministry of Land, Infrastructure and Transport, held an open call for citizen's ideas on urban planning for Sejong 5–1 Living Area using "Cities: Skylines." It is an example that shows how this video game imagery is uncritically adopted in governmental and popular understandings of the smart city. Image source: https://store.steampowered.com/app/255710/Cities_Skylines/.

context of smart city export between South Korea and Vietnam, these aspirations for transcending the physical limits of ordinary urban spaces can be linked to the desire for leapfrogging, to overcome their economic "backwardness" and finally catch up with the pace of up-to-date high technology and information-based developments.

Clearing land and starting from scratch has been practiced in South Korea and Vietnam before the era of smartness. In South Korea and Vietnam, physical redesign of the environment has been an ordinary practice since the twentieth century, especially since the postwar restoration, rapid industrialization, and economic development of both societies. At the same time, around the late 1980s, both countries entered the era of globalization as postcolonial nation states that had undergone industrialization following mid-twentieth-century wars. This process occurred more than a century after Europe and the US experienced industrialization.

An example of sociocultural aspirations behind erasing land is the demolition of the Japanese General Government Building in the heart of Seoul, South Korea. Built in 1910 as the Japanese colonial government's headquarters, the building was built right in front of Gyeongbokgung Palace, the main royal palace of the Joseon Dynasty built in 1395. Even after the end of the Japanese colonial occupation in 1945, the building remained until 1995. There's a saying that the Japanese government wanted to disassemble the building to preserve the parts as historical relics. However, then-president of South Korea Kim Young-sam decided that the building should be bombed and demolished completely to stage a kind of performance of the destruction of colonial structures in Korean society that had been blocking the "spirit of Korean people," symbolized by the Gyeongbokgung Palace. With that decision, the colonial building and the histories it represented were reduced to rubble. In this case, clearing land was not just about redesigning the physical landscape but a performative and rhetorical gesture to move on from the past and toward the digitalization of society.

Likewise, with the smart Thủ Đức City, geophysical and social environments are formatted and rebooted for the digital transformation of society that comes after demolition and eviction. As these cases demonstrate, land clearing has been justified by the notion of progress, and the latest form of progress is smartness: economic prosperity through technological innovation. Considering these specific histories of the rebooting and clearing of land reveals disparate contexts for the seemingly universal, developmentalist ideals of smartification through technological innovation. These differing contexts explain why smartification is adopted and spread at a faster pace in some countries than in others, and they show that smartification involves much broader social and cultural processes than technical ones.

In addition to the technical requirements for hardware and infrastructures to be installed and airwaves to travel with lower latency, cultural aspects are another element that facilitate the practice of erasing land. Building societies from scratch on emptied land has been a familiar practice in the history of South Korea and Vietnam. Their land has gone through constant deconstruction and reconstruction – global superpowers waged wars in the first half of the twentieth century on their territories, followed by the postwar and postcolonial construction of societies and economies. Due to this repeated mass-scale land clearing that South Korea and Vietnam had to cope with, it is common for the people and the government to be amenable to land clearing practices. Their open attitude applies to accepting trends of digitalization of everyday settings, which again affects the pace and degree of smartification of the physical world. Smartness is favored and adopted for developmental aspirations embedded in its sites of implementation. Geophysical redesign for smartness therefore is imbued with symbolic and cultural meanings that are specific to geographic contexts.

Historical Patterns of Cleared Land

Despite its acceptance in South Korea and Vietnam, it should be noted that land clearing causes large-scale destruction and involves two groups of people: those who order the destruction and those who bear its consequences. The seemingly empty, cleared land fully contains the complex histories around conflicting material conditions and aspirations that led to creating a massive piece of cleared land, and this practice can be situated in line with historical patterns involving land clearing.

Clearing land is a historical practice necessary for the remote use of power. Vietnam's land was a significant site of the development of aerial technologies by the US military – Operation Igloo White by the US Air Force was first executed in 1968 and dropped electronic sensors in the jungle along the Ho Chi Minh Trail to remotely detect the movements of enemy troops. The information gathered by the sensors was sent to the Infiltration Surveillance Center to determine targets for attack (Edwards 1996, 3). As one of the first sensor networks and a historical precedent for smart technologies, Operation Igloo White signifies the impulse to manipulate and rearrange the geophysical world for the remote-controlled, automated operation of the world through a computing system. (A similar impulse unfolded in the use of a different technology, chemical defoliants such as Agent Orange, to damage and remove things on the land.) Paul Edwards considers Operation Igloo White as an instance of techno-strategic developments playing out on a regional scale (Edwards 1996, 6). Edwards argues that it allowed its system operator to achieve practical goals – whether it was the repressive power of surveillance and bombing or the productive power of remote sensing techniques that achieved an advantageous position in the war and Cold War dynamics (Edwards 1996, 40).

How is this impulse in Operation Igloo White different from the reasoning behind land clearing in Thủ Đức City in the 2010 and 2020s? Anthropologist Erik Harms (2016), who has investigated land rights and eviction issues in Ho Chi Minh City, demonstrated the dialectics of urban development in which "luxury" (fancy, modernized, newly built housing and residential area) and "rubble" (demolished houses and evictees' lives that were part of the project of clearing land in the Thủ Đức City construction site) go hand in hand in his book *Luxury and Rubble: Civility and Dispossession in the New Saigon*. Harms (2016) argues that the "rubble," the demolished houses, represents the vernacular ways of life that had been built in line with the postwar recovery and the condensed pace of industrialization that culminated in the "luxury" that has been a big part of modern and contemporary Vietnamese life. Harms's discussion points to the dialectic co-construction between "luxury" and "rubble" within the context of the post-reform era of Vietnam in its transition to market-oriented socialism (2016, 2–7). In this chapter, I situate the same site and practices on an expanded field of post-industrial, global capitalism. This renewed scope allows for identifying similar historical patterns of removing pre-existing lives and objects on the ground while at the same time tracing the bigger cause that justified the destruction.

These "rubble" houses block Vietnam's path to realizing the 5G communication needed for a smooth transition to digitalized society. These houses also get in the way of creating an urban zone that showcases Vietnam's ability to implement technologies that live up to the standards of Silicon Valley. Clearing land in this case aimed to remove things that may get in the way of Vietnam's digitalization and to leapfrog ahead through technological innovation. This resembles the way land in Vietnam was treated in Operation Igloo White, an electronic warfare in which sensors were planted on the ground to generate networks and communicate with aircraft. As Gabrys and Edwards observe, this form of networked electronic warfare is about the environment and how the merging of environment and computing produces a conduit for power dynamics and economic interests (Gabrys 2016; Edwards 1996). Through land clearing in the guise of smartness, global (in other words, foreign or multinational) enterprises conduct extraction and control of capital and data, making power plays among different actors across the land of South Korea and Vietnam.

Smartness can be understood as a form of mediated space, a spatialized system that ultimately serves the aspirations of latecomer countries for economic prosperity through digital technologies, aspirations that are intertwined with a familiar historical pattern of power play and capitalist intervention. What is important to note is that geophysical and ideological dimensions of space always go hand in hand. That is why void space – the outcome of land clearing discussed in this chapter – comes to the fore as a prerequisite and common ground for smartness when its virtual and spatial envisioning is put into a transnational context.

Smartness as an Excuse for [Fill in the Blank]

In this chapter, we have broadened the scope of smartness and its infrastructural system by considering land-clearing practices as their material precondition. Starting from the technical aspect of turning the physical into the digital, we have considered the social aspect of smartness that supports the national-level economic agenda and digital innovation initiatives. By putting the national-level aspirations into a transnational context, we have examined that smartness is not universal but necessary for context-specific purposes. Zooming out even further on a temporal scale allows for identifying a more historical pattern regarding erasing land – we can observe a curious parallel between erasing land using herbicide during the Vietnam War and erasing land using eviction and bulldozers in constructing a smart environment in Thủ Đức City.

Thinking about smartness in relation to the land, which is being erased repeatedly across time and place, allows us to understand that smartness can be a surface level excuse that justifies resurfaced forms of invasion and occupation by foreign multinational powers. Similar to the way colonial and Cold War invasions during the industrial era began by occupying the land, in the era of post-industrial digital economy, invasion and occupation happen on the erased land: cleared land not only provides an ideal setup for context-free, scalable digital technology, but it further paves the way for transnational flows of money and power.

Both South Korea and Vietnam served as the backdrop for wars by global superpowers in the twentieth century that were marked by many forms of land clearing. Now, they have positioned themselves among the world's most digitally competitive and innovative economies, having emerged from the rubble of those wars. Leveraging their positions as latecomers in industrial society, South Korea and Vietnam strive to catch up and excel in a post-industrial, digital-based economy. These aspirations manifest in the form of cleared land, which is a crucial yet often underrecognized infrastructure of smart cities. Clearing the land not only facilitates the connection between the computer system and the things in the physical world, it is also the standard practice in these societies to eliminate material and symbolic barriers obstructing innovation. Still, what is ultimately mediated in this (clearing of) geophysical land is the same old invasive, capitalist power play by foreign forces. Smartness is rooted in cleared land.

References

Edwards, Paul N. 1996. *The Closed World: Computers and the Politics of Discourse in Cold War America.* MIT Press. https://doi.org/10.7551/mitpress/1871.001.0001.

Gabrys, Jennifer. 2016. *Program Earth: Environmental Sensing Technology and the Making of a Computational Planet.* University of Minnesota Press. https://doi.org/10.5749/minnesota/9780816693122.001.0001.

Halpern, Orit, Jesse LeCavalier, Calvillo Nerea, and Wolfgang Pietsch. 2013. "Test-Bed Urbanism." *Public Culture* 25 (2): 272–306. https://doi.org/10.1215/08992363-2020602.

Harms, Erik. 2016. *Luxury and Rubble: Civility and Dispossession in the New Saigon.* University of California Press. https://doi.org/10.1525/luminos.20.

Im, Jae Yoon, and Hyungsub Choi. 2017. "Choi Hyung Sup and the Origins of the Korean Development Model (최형섭과 한국형 발전 모델의 기원)." *Critical Review of History* 118 (February): 169–93. https://doi.org/10.38080/crh.2017.02.118.169.

Lee, Tae Joo. 2021. "An Anthropological Reflection on the 'Korean Style' Development Assistance Culture: With a Focus on the Global Saemaul ODA and the KSP (Knowledge Sharing Program) ('한국형' 개발원조 문화의 인류학적 성찰: 글로벌 새마을 ODA와 경제발전경험공유사업(KSP)을 중심으로)." *Korean Cultural Anthropology* 54, no. 3 (November): 43–84. https://doi.org/10.22913/KOANTHRO.2021.11.30.3.43.

Ministry of Culture, Sports and Tourism of South Korea. 2021. "Policy Wiki: New Southern Policy." September 6. www.korea.kr/special/policyCurationView.do?newsId=148853887.

Ministry of Land, Infrastructure and Transport of South Korea. 2021. "Smart City Korea." April 28. https://smartcity.go.kr/2021/04/28/스마트시티-해외진출-전략-보고서-k-city-network-2020-스마트시티/.

Thế, Lâm. 2020. "Thu Duc Is Prioritized for 5G Coverage to Build Digital Government (Thủ Đức được ưu tiên phủ sóng 5G để xây dựng chính quyền số)." *Lao Động,* December 31. https://laodong.vn/xa-hoi/thu-duc-duoc-uu-tien-phu-song-5g-de-xay-dung-chinh-quyen-so-866928.ldo.

PART TWO

Plastering Frictions and Fissures with Smartness

7 Introduction to Plastering Frictions and Fissures with Smartness

VINCENT MIRZA

For smartification (and capitalism) to function, it needs to be based on a normative framework that standardizes procedures that can be measured and replicated. In contrast and in the real world, part 2, "Plastering Frictions and Fissures with Smartness," examines how the tension between this normativity and heterogeneity create what Anna Tsing (2005) calls frictions. Frictions emerge where dominant views of the world translate it into technological forms of imposition but encounter forms that escape this frame, forms that are more complex to catch and standardize. Here, the battlegrounds of frictions are an attempt to impose both a definition of what the world should be, and categories of thought through which we ought to think about smartness.

Indeed, thinking with Tsing, frictions are moments of tension, the space in which various forms of understanding and thinking of the world come to be negotiated, generating conflicts but also solutions and cooperations. What the frictions reveal is that power relations are at play in a network of representation and mode of existence that bring to life new dynamics. In part 2, it is interesting and productive to think of categories, such as smart cities (Mu) or smart buildings (Dorthe and Dobigny), as sites of frictions because they remind us of this tension between a messy and unpredictable world, and the articulation of technoscientific smartness that needs to work on repetition, data, and predictability. Frictions help us think about how data articulates and imposes a relationship upon life and space, and how we operate within it.

To paraphrase Tsing (2005), the notion of friction reminds us that it is the "futures" that are in the balance, that zones of friction are often the articulation point between assemblages of a dominant perception of the world imposed through technologies – that like capitalism organize categories and content – and a world that escapes these logics and produces alternate possibilities of life. This is where fissures are offering an alternate space where unplanned, messy contradictions can emerge, those that escape the standardization and predictability of data and smartification. This is also where a political discourse or the performative narrative of capitalism tends to create an illusion of a smooth smart future and where the reality of frictions shows us instead that it creates something new. Indeed, the notion of friction for Tsing is not only about creating forces and making new or unexpected connections. The notion of friction adds complexity to explanatory schemes that would be based solely on growth and reproduction, and it insists on the idea that interactions give rise to something new, co-produced by the meeting of seemingly disparate flows.

Contributors in part 2 bring to the fore and shed light on how space and narratives articulate together in an attempt to produce a world without friction and fissures. They also reveal how material realities and postcolonial futures (Junnan Mu) are obscured by the performative aspects of smartification and smart cities. Gabriel Dorthe and Laure Dorigny, in examining a smart building, also reveal the contradiction – frictions – between a smart technology and smart users. Simon Rabyniuk brings to the fore the question of simplification and smartness that create an illusion. Finally, this section is concluded by Sarah Marquis, who reveals through the study of digital agriculture how smartness is sometimes conveniently conflated with the notion of sustainability. Through these contributions, we are reminded that frictions, fissures, and repairs come into play to create a reality that needs to be considered at the intersection of smartification and life.

Reference

Tsing, Anna Lowenhaupt. 2005. *Friction: An Ethnography of Global Connection.* Princeton University Press. https://doi.org/10.1515/9780691263526.

8 How Not to Define the Smart City

JUNNAN MU

As we passed through the southeastern peri-urban areas of Nairobi on our way to Kenya's first smart city, government officials responsible for implementing the smart city project frequently drew my attention to the "informal settlements" in the vicinity. The industrial and commercial development in between the capital city and the smart city has led to the expansion of peri-urban spaces to accommodate rural-urban migrants. However, the peri-urban landscape is not extensive. Beyond the town of Athi River, an endless savannah landscape unfolds, where herds of sheep and goats graze freely. Along the route, the proposed Konza Technopolis, a smart city governed by the Kenyan state, can only be identified by a green billboard on its border displaying the words "Silicon Savannah." A solitary structure framed in blue glass windows, stands in stark contrast to its rural surroundings, mirroring the architectural style prevalent in the numerous skyscrapers that punctuate Nairobi's central business district. For those residing near the Konza Technopolis, the smart city is referred to by a more practical name, "the ICT land," indicating its ownership by the Ministry of Information, Communications, and Technology (MoICT). While the absence of the term "smart city" among those living in its hinterland does not necessarily contradict the officially subscribed significance of this technologically driven urban ideal, the multiplicity of what "smartness" denotes locally make it a rather contingent idea.

In this chapter, I explore the "definitional impasse" of the "smart city," arguing that practices on the ground and the lack of coherent meanings tap into the common yearning for a different postcolonial urban future. By understanding the definitional impasse, we can gain insights into the process of smartification in an African locale that involves the simultaneous transition to urbanization and digitalization. As a flagship project of Kenya Vision 2030, the Konza Technopolis represents a new phase of the digital future of Kenya. It is envisioned to catalyse the development of Kenya's knowledge economy and innovation. Thus far, the trajectory of digital development in Kenya has proven to be unpredictable and disruptive. Despite the absence of significant investment, Kenya achieved groundbreaking innovation in mobile payment (M-PESA). The use of M-PESA, together with ongoing government efforts to expand fibre-optic networks and the emergence of private tech hubs and innovation centres, positioned the country as a leading destination for digital innovation and investments. Therefore, Kenya's encounter with digital technology encompasses more than just economic progress; it also lets ordinary Kenyans visualize their near future based on their present circumstances. The making and remaking of a smart city in Kenya thus offers us a lens through which to see a potential new phase in the history of technology and development in the Global South, where grand promises, even when unfulfilled, catalyse numerous enterprises on the ground to produce something new.

Rather than seeking a concrete definition of the smart city in the Kenyan context, I explore instead how a definitional impasse is manifesting itself on the ground, particularly within the institutional space of the Konza Technopolis Development Authority (KoTDA), a state corporation under the Kenyan MoICT. Similar to numerous future-making projects in the Global South, new town developments labelled as "smart cities" often involve local investments in the performative efficacy of an abstract ideal that lacks a fundamental material basis. Kenyan technocrats and bureaucrats are very much aware of such discrepancies, and they mobilize different semiotic strategies to fix the meanings of "smart city." Yet it is from those multi-level efforts of meaning-fixing that the definitional impasse surfaces.

Anthropological theories and methodologies are particularly proficient in unpacking the fissures enacted by global projects and ideals thanks to the discipline's sensitivity to the connection and disjunction of social processes and practices at different scales (Ong and Collier 2005; Tsing 2005). Specifically, I propose using the *making* of Konza Technopolis as the generative analytical and methodological framework for us to understand the actual processes and dynamics discharged by the smart city ideal. Implementation represents the liminal stage that exists between aspirations and realities, hopes and doubts, and more tangibly, between a utopian promise and its dynamic adoption on the ground. In anthropological theory, liminality signifies a "threshold" and implies the spatiotemporal reconfiguration of transformative potentials that often surpass the confines of fixed social categories (Turner 1967). By applying this concept to the broader socioeconomic transition in Kenya, I argue that the definitional impasse reflects local agency that challenges the universal imaginary of the smart city.

Amid the interplay between a universalist urban ideal and localized frictions, figurations and narratives of the smart city become contested, giving rise to creative potentials that disrupt the rigid linearity imposed by master planning. It is important to note that foregrounding the definitional impasse of the smart city does not imply rejecting the smart city's ideological affinity with neoliberalism and techno-utopianism. As I will demonstrate in the subsequent sections, by not defining the "smart city" a priori, a potent interpretative community focusing on the cosmopolitan origins of the ideology and implementation of the smart city emerges, undermining its technological underpinning and resocializing its marketing value for IT companies and difference levels of government. I conducted a three-month institutional ethnography within the KoTDA from February to May 2019. I followed speech events (meetings, public speeches, and workshops) that evoked the concept of the smart city and conducted interviews with officials, board members, and subcontractors. I also conducted focus group discussions in villages around Konza Technopolis. In the following, I will first provide a concise historical overview of Konza Technopolis. Then, I will delve into the negotiations and tensions that surfaced during the implementation phase within and between the KoTDA and the local community. In conclusion, I advocate for a more nuanced comprehension of the symbolic excesses of the "smart city" and its constitutive social processes.

Planned and Unplanned Smartness

In the past decades, different strands of urban aspirations have converged under the rubric of the "smart city." The pursuit of smartness is unprecedentedly associated with the mechanical and artificial engineering of intelligence in social systems beyond the human. As both an urban ideal and a pragmatic process, the smart city presupposes that the social can be engineered and governed by allocating authority to the data and machine (Mattern 2021). In Kenya, the national economic plan – Kenya Vision 2030 – has brought many infrastructural projects into its developmental horizon. According to Kenya Vision 2030, Konza Technopolis is the engine of national economic growth. For instance, on February 14, 2012, the Kenyan newspaper *Daily Nation* spent a whole page to announce the coming of Konza Technopolis as the emblem of "time for more Kenyans to venture into cyberspace" (Ongwae 2012). Alongside the Kenya coat of arms and the logo of Kenya Vision 2030, a 3D-rendered image of glowing skyscrapers and boulevards with an excess of walking space created the initial impression of this smart city. Under such a vision, the development of a master-planned smart city and associated infrastructures dubbed the "Silicon Savannah" has provided a legible orientation towards a more vibrant digital economy to enhance Kenya's role as the tech hub in the continent. Although many African countries adopt the smart city ideal in their urban policies (mainly in the form of new town development), smart city projects that have emerged across the continent have been criticized as "urban fantasies" (Watson 2015, 2014). Critical urban theories focusing on neoliberalist space-making have emphasized socio-spatial marginalization, exclusion, and alienation as effects of the rupture between universal "planning" and actual social conditions (Brenner 2003; Graham 2001). They offer a comprehensive understanding of the broader societal impact of such projects, particularly in relation to the privatization of public space and the exacerbation of social segregation (Datta 2015; Watson 2014). While I am sympathetic to those critiques, I think the association of the smart city with the neoliberalist "urban fantasy" too quickly ignores active imaginaries and future-making practices that have emerged on the ground. In this context, it becomes crucial to thoroughly understand the actual process of creating a smart city and the historical, social, and economic conditions that underpin the development of these imaginaries and practices.

Kenya's modernization paradigm has undergone a significant shift in relation to the enhanced capacity of digital technologies, which set the stage for the emergence of the Konza Technopolis. In 1996, influenced by the World Bank's "Information and Communication Technologies for Development" (ICT4D) framework, the United Nations Economic Commission for Africa established the Africa Information Society Initiative (AISI) to foster information and communication technology infrastructure on the continent. It's worth noting that ICT4D was not the only developmental framework that gained traction in Africa, and Kenya was not the sole country strongly influenced by it. However, what distinguishes Kenya is its stable ICT policy and supportive environment for scaling up innovation. As an active participant in the AISI plan, the Kenyan government developed its own vision and strategy for national information and communication infrastructures.

Notably, AISI promoted neoliberal ideals that aimed to reduce the government's role as a service provider (Ojo 2016). Since 2004, Kenya has pursued the privatization of its ICT industry, resulting in the removal of the state's monopoly over telecommunications, radio, TV, and internet services. This move has allowed the private sector to provide essential operations in these sectors. Accompanied by successful fibre-optic infrastructure development and the widespread adoption of mobile phones, Kenya's ICT industry has experienced significant growth and innovations (e.g., M-PESA), offering a competitive market for technology-related investments. This growth has bolstered the government's confidence in identifying Business Process Offshoring (BPO) and IT-Enabled Services as two of the six drivers of the economy outlined in Kenya Vision 2030. These two drivers later became the underlying rationales for the conceptualization of Konza Technopolis. In 2013, President Kibaki issued the Konza Technopolis Development Authority Order, establishing the KoTDA to oversee the implementation of the Konza Technopolis project. As the first and only state-led smart city initiative in Kenya, Konza Technopolis relies on both multilateral collaborations (involving the governments of China, Korea, and Italy, among others) and public-private partnerships as new policy frameworks and financial mechanisms to advance its development. Consequently, the bureaucratic mediation of these public-public and public-private interests deeply shapes the discursive practices within the KoTDA.

During my fieldwork, I was often impressed by metaphors and communicative strategies used by KoTDA's officials when they explained what they meant by "smart city." In practice, officials used the chaotic image of Nairobi to justify the value of a master-planned city. For example, local imaginaries of a smart city are constituted by the negation of whatever uncomfortable urban life the city of Nairobi has afforded, such as "informal settlements," congested roads, and the untenable roofs. Ideally, the smart city in Kenya is a new city planned through mixed-use development. It is well designed and guarantees the order, safety, and sustainable life that Nairobi cannot offer. Utopic as it is, Konza Technopolis is their promised land, invested with the ideas of their common futures in a new space of prosperity. An official from KoTDA once explained to me the rationale behind the absence of affordable housing in the planning of the Konza Technopolis. "All the technology will come from Konza, we might not have affordable housing, but we will generate technology for constructing affordable housing." At first glance, the official's response seems to suggest that Konza is the production site of technology. However, by affirming the ideal of spatial segregation, Kenyan official reinforces the idea that technology can only originate from master-planned, well-designed, and often elite zones, obscuring the social fabrics, that is, the labour, local resources, and spatial inequalities that facilitate the production of technology.

In contrast to the master plan, in nearby villages around Konza Technopolis, a form of symbolic currency took shape through villagers' use of the terms "cyber," "tech," and "digital" to name their businesses. Many villagers established businesses such as grocery shops, restaurants, and stationery shops to capture the nevertheless temporary economic opportunities that Konza Technopolis generated in the meantime (mostly from construction workers living nearby). Although the term "smart city" was significantly absent from people's everyday communication, the use of "cyber," "tech," and "digital," along with officials' profligate conjuration of hope and utopianism, contributed to the efficacy of the "smart city" as the symbolic anchor of future-making that could be shared and appropriated by different social groups. Moore and Smith (2020) have observed that ordinary Kenyans creatively mobilize the term "digital" as a signifier of their experience of rapid social changes and shifting perceptions of time. By equating "smartness" with enhanced digital capacity, these local expressions of the ambiguous "smart city" inadvertently affirm the ideology of techno-solutionism. Furthermore, as the project's implementation has progressed slowly, the technological justification for the state's promotion of the smart city has found validation through the local use of these buzzwords. The collective yearning for utopia, or even just a more viable mode of existence, resonates with demands for remaking Kenya in the new political economy of digital capitalism, characterized by increasing technologically facilitated connections and investment. In this process, "smart city" becomes an empty signifier of the everyday possibility of beneficial changes. It diverges from the economic promises focusing on the IT-driven services and BPO outlined in Kenya Vision 2030 and finds more convenient and immediate grounds on economic forms that seldom involve any digital means. For the development authority, those local representations of the smart city not only reveal the pervasive grassroots fantasy about digital technologies but also challenge the state-envisioned material basis of "smartness" exemplified by large-scale physical infrastructure projects. In fact, the increasing land speculation and "unplanned" development around the smart city site was considered by KoTDA officials to be a threat to the "integrity" of Konza Technopolis.

One argument commonly employed by KoTDA in the public sphere is that of cosmopolitanism. According to the development authority, "smart people" are a population qualified with "Social and ethnic plurality, Creativity, Cosmopolitanism and Lifelong learning" (KoTDA 2019). Such a qualification directly refers to the European smart city model developed by researchers such as Rudolf Giffinger and his research team at the Technical University Vienna in 2007 (Giffinger et al. 2010). Based on the European smart city model, "smart people" is one of the six key fields of urban development. Compared with the European models, KoTDA added the "cosmopolitanism" ethics to the "smart people" field.[1]

---

[1] The definition of smart city has seen modifications from 2019, which again proves that constant efforts are made in defining, redefining, and undefining the "smart city."

A long-debated dilemma of cosmopolitanism emerges concerning how such conviviality can be integrated into the local values of belonging. However, cosmopolitanism does allow for the creative and open blending of techniques from different points in the developmental process. In general, the Kenyan government perceives "smartness" as a model of urban governance and the new frontier of the digital economy that has been conceptualized in wealthy and technologically advanced countries. Such a particularistic portrait of "smart people" is embedded into KoTDA's officials' aspirations. Nevertheless, despite reproducing the image of the global elite community, the cosmopolitan narrative also highlights powerful interpretations and creates space for the mutual transformation of "smart people" and the smart city. KoTDA's reworking of the "smart people" resonates with the aspirations of young people. The promise of intermingling with a prominent "cosmopolitan" (multinational and multi-ethnic) community attracts a lot of local graduates to work with KoTDA in a time when "cosmopolitan" becomes the symbolic capital for people seeking upward mobility. In making a smart city, KoTDA's staff and local villagers are also aligning themselves to become appropriate residents of the place. Furthermore, the concept of "smart people" also promotes a standardized notion of economic potential and capital, inadvertently leading to exclusions and displacements by upholding a sanctioned image of expertise.

During an interview, a KoTDA official used the term "urban acupuncture" to describe how the the technopolis would generate knowledge, cultivate human capital, improve the economy, and directly impact people's way of life. "Urban acupuncture" refers to addressing minor issues that can have positive structural effects on the entire urban system. It perceives the city as an interconnected whole, where specific interventions can have broad-reaching impacts. The usage of this term by my interlocutor seemed to leverage its cultural connotations, particularly to resonate with a Chinese researcher. Indeed, Kenyan engineers in charge of building the smart city embrace technologies of various epochs and locations. What is more significant is the Kenyan planners' dedicated efforts to construct a plausible image of a "smart city" that accommodates diverse and, at times, contradictory interests. In fact, compared to other smart city projects in the Global North, Konza Technopolis boasts a notable multicultural profile. Through various financial mechanisms, the Kenyan government has attracted a range of international partners to the project, including American architects, Italian technicians, Korean engineers, Indian managers, Chinese programmers, and many others. However, the pervasive presence of prominent foreign investors also instils a sense of unease regarding the true source of digital empowerment.

While many young Kenyans actively utilize mobile phones and the internet to navigate the uncertainties of their everyday lives and make connections with the outside world, the prolonged anticipation for the realization of the grandiose "Silicon Savannah" often breeds uneasiness. Consequently, many villagers find themselves lowering their expectations as a means of self-protection against the disappointment of unfulfilled promises. On one hand, there are ambivalent assumptions about the smart city as the panacea to the everyday precarity of postcolonial economy. On the other hand, no one is certain about what a smart city really is, yet they still act as if it is real. In the end, even the most sentimental evocation of the smart city bears the generative "semiotic force of alterity" (Newell 2012). Observed among Ivorian youth wearing extravagant fashions, "alterity," in Newell's characterization, is a local claim to superiority and difference that neither involves an "eye-winking slight of colonial whiteness" nor attempts to achieve postcolonial authenticity. While officials assume cosmopolitan citizenship as the basis of the smart city, local villagers act through improvisation and mimesis to realize their version of this techno-driven postcolonial urban imaginary, characterized by the juxtaposition of the "digital" outside and the local inside. Therefore, no matter whether the claimed "smartness" is factual or fictional, it paves the way for whoever evokes this concept enact its performative efficacy and, thus, be incorporated in the broader global digital economy.

Last, the definitional impasse also serves to legitimize the practical impasse in terms of the project's unsatisfying progress. KoTDA strategically mystifies the new tech of a smart city to ask for more patience in response to critiques on its slow implementation. As local planners and engineers grapple with the width of the roads, the size of the land, and other material aspects to align them with the criteria of "smartness," they also face the challenge of justifying their decisions to other bureaucrats. Defining and undefining a "smart city" becomes an everyday puzzle that requires decisive action.

Conclusion

In this chapter, I suggest that the definitional impasse of the "smart city" in Kenya is more than an unintentional effect of the slow implementation of an actual state-led project. Despite many efforts being made to *define* and thus to create semiotic coherence for the smart city, the impasse arises not as a result of conflicting meanings but is constituted from the residue of unattainable yet hopeful future upon which many are relying to revitalize the fragmented livelihoods in the former colony.

In other words, the definitional impasse is symptomatic of the smart city being an empty and floating signifier that always has to seek local expression. In Kenya, this empty signifier is isomorphic to the expansive empty landscapes both within and beyond Konza, fetishized by local aspirations and global speculations.

Taking this issue further, the temporality of smartification, despite being structured by master planning, is a speculative one and is contingent upon local socioeconomic configurations. As shown above, long before any signs of development have manifested in the ICT site, the symbolic valence of the "digital" creates an alternative community in contrast to the cosmopolitan "smart people." For both the state and the local community, the "smart city" is a practice to hedge uncertain futures. If not non-linear, the actual process of smartification is at least not blocked in the timeline set by state planning documents or the projects' delivery schedule. As shown in local businesses' appropriations of signifiers such as "cyber" and "digital" around Konza Technopolis, the state's vision of "smart city" does not precede the emergence of those local discursive practices from creative uptakes of the promise offered by digital tools and means of life. Meanwhile, both government officials and local villagers illustrate opportunistic clarifications of "smart city," pulling the term away from the planning mechanism underpinning its initial justification. Through the elusive pursuit of digital novelty, different forms of collectivity take place. Vulnerable and unstable, they continue adding scripts to dissolve the solidity of the definition of the "smart city," making it an always transient assemblage of different ideologies, social institutions, and cultural norms that structurally constitute the desire for novel signifiers of the good life.

Furthermore, through understanding the definitional impasse of "smart city" in the case of Konza Technopolis in Kenya, I echo Günel's (2019) insights in moving beyond the binary critiques of the "failure" or "success" of a smart city. Both government officials' opportunistic interpretations and the marginalized local community's hopeful beliefs are not just the ramifications of techno-solutionism or techno-utopianism. If anything, "technology" here mainly serves to enhance the vernacular impression of modernity. Indeed, the process of creating a smart city in Kenya, such as Konza Technopolis, reflects the longstanding strategies employed by many Kenyans to navigate and succeed in exploitative and competitive urban environments. Rather than adhering to a rigid definition of what a smart city should be, local actors in Kenya adopt flexible and adaptive approaches characterized by "cosmopolitanism" (embracing diverse influences). These approaches are crucial for individuals to shape their own identities and envision their future in a postcolonial society. Meanwhile, this also means a degree of negation of existing lived experiences in Nairobi as well as the acknowledgment and creative navigations of the potential of change the smart city enables. I suggest it is meaningful to bring "smart city" to the ground, to its actual making, not merely for contextualization, but to unpack the constituents of the elusiveness of an ideal circulating around the globe. In this regard, it is critical to take "smart city" not simply as the technological fix of capitalist regimes of different scales, speeds, and intensities but as the assemblage of intersubjective values produced in multiple social, semiotic, and material processes as lived on the ground.

References

Brenner, Neil. 2003. "'Glocalization' as a State Spatial Strategy: Urban Entrepreneurialism and the New Politics of Uneven Development in Western Europe." In *Remaking the Global Economy: Economic-Geographical Perspectives*, edited by Jamie Peck, Henry Wai-chung Yeung, and Neil Brenner. Sage. https://doi.org/10.4135/9781446216767.n12.

Datta, Ayona. 2015. "New Urban Utopias of Postcolonial India: 'Entrepreneurial Urbanization' in Dholera Smart City, Gujarat." *Dialogues in Human Geography* 5 (1): 3–22. https://doi.org/10.1177/2043820614565748.

Giffinger, Rudolf, Gudrun Haindlmaier, and Hans Kramar. 2010. "The Role of Rankings in Growing City Competition." *Urban Research & Practice* 3 (3): 299–312. https://doi.org/10.1080/17535069.2010.524420.

Graham, Stephen. 2001. *Splintering Urbanism: Networked Infrastructures, Technological Mobilities and the Urban Condition*. Routledge. https://doi.org/10.4324/9780203452202.

Günel, Gökçe. 2019. *Spaceship in the Desert: Energy, Climate Change, and Urban Design in Abu Dhabi*. Duke University Press. https://doi.org/10.1215/9781478002406.

KoTDA. 2019. "Introduction to Konza Technopolis: Infrastructure and Services (2019-02-11)." PowerPoint Presentation.

Mattern, Shannon. 2021. *A City Is Not a Computer: Other Urban Intelligences*. Princeton University Press. https://doi.org/10.1515/9780691226750.

Moore, Henrietta L., and Constance Smith. 2020. "The Dotcom and the Digital: Time and Imagination in Kenya." *Public Culture* 32 (3): 513–38. https://doi.org/10.1215/08992363-8358698.

Newell, Sasha. 2012. *The Modernity Bluff: Crime, Consumption, and Citizenship in Côte d'Ivoire*. The University of Chicago Press. https://doi.org/10.7208/chicago/9780226575216.001.0001.

Ojo, Tokunbo. 2016. "Global Agenda and ICT4D in Africa: Constraints of Localizing 'Universal Norm.'" *Telecommunications Policy* 40 (7): 704–13. https://doi.org/10.1016/j.telpol.2016.05.002.

Ong, Aihwa, and Stephen J. Collier, eds. 2005. *Global Assemblages: Technology, Politics, and Ethics as Anthropological Problems.* Blackwell.

Ongwae, Evans. 2012. "Kenya to Break Ground for First Smart City." *Daily Nation,* February 7.

Tsing, Anna Lowenhaupt. 2005. *Friction: An Ethnography of Global Connection.* Princeton University Press. https://doi.org/10.1515/9780691263526.

Turner, Victor W. (Victor Witter). 1967. *The Forest of Symbols: Aspects of Ndembu Ritual.* Cornell University Press.

Watson, Vanessa. 2014. "African Urban Fantasies: Dreams or Nightmares?" *Environment & Urbanization* 26 (1): 215–31. https://doi.org/10.1177/0956247813513705.

— 2015. "The Allure of 'Smart City' Rhetoric: India and Africa." *Dialogues in Human Geography* 5 (1): 36–9. https://doi.org/10.1177/2043820614565868.

9 From Smart Buildings to Smart Users? Energy Transition to the Test of Parasitic Humans

GABRIEL DORTHE AND LAURE DOBIGNY

Introduction

The Live TREE Program[1] at Lille Catholic University in France clusters various initiatives that share the aim of enabling ecological and societal transition in all dimensions of campus life. More broadly, the program explores, trials, and demonstrates the feasibility of desirable futures in times of growing ecological devastation. Funded by the Region Hauts-de-France, it is meant as a local response to national and international commitments, such as the Paris Agreement (UNFCCC 2015). As of today, the flagship achievements of Live TREE are the renovation of two historic edifices as "smart buildings" for hosting research, teaching, and administrative activities. They are meant as "demonstrators" of how sensors and IT systems coupled with renewable energy can and should pave the way for a less carbon-intensive future and more sustainable ways of life.

In February 2020, a semi-annual meeting took place at the control centre of the program, where data is processed to monitor and adjust smart building parameters. Amid a highly technical conversation, someone pointed out the need for "smart users" to inhabit these buildings that were still (and still are) undergoing trials and adjustments. Embarrassed, the chair of the meeting nodded at one of us, sitting in on the meeting as an invited observer, and said: "Well, here, we could benefit from the help of ethicists, as we are navigating uncharted territory." Having put so much effort into renovation and equipment, they were animated by mixed feelings of amusement and concern, wondering if they had to

---

1 This acronym stands for "Lille Vauban en Transition Énergétique, Écologique et Économique" (Lille Vauban in Energy, Ecology and Economic Transition). We are immensely grateful to Damien Bright, who helped us improve the English of this chapter.

craft appropriately fitting humans for their building. Another engineer whispered: "Otherwise, we might just end up with another fancy building stuffed with sensors!" What is the expected standing of facility users within this emerging paradigm? To what extent can they play an active role in this technical approach to environmental issues, where energy efficiency is considered the main solution?

Despite its breadth of aspirations and outspoken concern for social dimensions, Live TREE is largely driven by technological innovation. It is meant as an implementation of the Third Industrial Revolution paradigm detailed by Jeremy Rifkin, which promotes sustainability through technology innovation by harnessing digital technology and renewable energy (Rifkin 2011). While energy transition and sustainability require drastic and multifaceted transformations – in infrastructure, policy, collective imaginaries, and individual behaviour – technological dimensions are often favoured over social ones in order to scale up and save time, as emergencies are on the rise and long-held habits seem difficult to shift.

Yet individual and collective practices have a strong tendency to complicate the smooth picture of transition that designers plan for, whether through passive or active resistance (Houston et al. 2019). Such complications, in turn, are a reservoir of imaginaries and practices that could help designers refine their processes and outlooks.

We are both, in various capacities, embedded researchers within Live TREE and participate in its collective undertaking. In this sense, our involvement in the program could be defined as participant observation (see e.g., Bernard 2018; Jorgensen 2020), that is, participating and being involved in a social organization, not only studying it as external observers. This puts us in the relatively privileged position of being able to see aspects and nuances of the situation that would be scarcely visible from the outside and also to actively contribute these observations to discussions. Yet that might also prove uncomfortable as we are perceived as fully fledged members of the group, sharing its values and perspectives (Dubey 2013). As STS scholars and social anthropologists of technology, our role is not to answer the preformatted questions that engineers tend to address us with, but to question their assumptions – as much as our own. The point is to change the framing of the questions, *together*, rather than to come up with answers. Scholarship on smart projects around the world, of which the current section of this book offers a stimulating illustration, also pushes us to question the broader picture of how an allegedly universal ideal – smart technology is a good solution to many issues – interacts with local particularities in multiple, and sometimes surprising, ways.

In this chapter, we explore two conceptions of the coproduction (Jasanoff 2004) of users and smart buildings that engineers tend to hold and then examine enduring sources of instability regarding the ways in which these buildings are expected to fulfil their objectives and meet users' needs. Are users threats or allies for a building's energy efficiency? When the latter is exclusively governed by data and technology, the building is meant to be completely autonomous, and users must remain passive to avoid interfering with its performance. When more attention is given to individual practices – and to how they can help to decrease energy consumption – the building remains more flexible and able to adapt to various situations and users. In this case, users are called to take a more active role in its "life."

Smart Building Imaginary: A Living Organism Parasitized by Users

"We need to see how it *breathes*." With the help of data and sensors, Live TREE engineers and technicians describe smart buildings as living organisms that "*breathe*," "*live*," and "*survive*."[2] Yet buildings are assisted by technical systems (for ventilation, for example), so their survival is far from being organic and depends heavily on maintenance and care. A strong imaginary of connectivity permeates smart buildings, between different systems, networks, and people. By using sensors that generate data and feed it to a centralized control centre, engineers produce a body of knowledge that is essential for the smart building to "regulate itself" or "alert" the technicians in case of irregular situations. This produces a singular relationship between technicians and smart buildings, a relationship in which buildings are black boxed and users tend to be kept away from its vital signs and organs. For example, users do not have access to the main settings (e.g., room ventilation or heating) and are not invited to promotional events (e.g., the inauguration of the building or public events) (Cherry et al. 2017; Garabuau-Moussaoui 2014).

Scholarship in social anthropology and STS has demonstrated the intrinsic non-neutrality of technology. Not only does it have social and political implications, by shaping social representations and practices it also embodies social dimensions, representations, and imaginaries (Winner 1986). Thus, technology has an impact on the social world as much as it is impacted by it (Gras 2003). Put slightly differently, technology and social norms are coproduced (Jasanoff 2004). Therefore, the uptake of technology and innovation is more than a matter of acceptability. It is closely connected to questions of worldview and social outlook, which in our case include the stakes of comfort, energy efficiency, desirable

---

2 Verbatim, Rizomm building visit, April 6, 2021.

futures, or the interplay of technology and users when tackling environmental issues.

While engineers think of smart buildings as serving users' comfort, the building's efficiency is completely cut off from its uses and users. Technology and user experience are envisioned separately. Hence, heterogeneous and sometimes conflicting performance goals get intertwined: energy efficiency, carbon neutrality, and a high level of user comfort (e.g., heating, ventilation, lighting, cooling, and connectivity). In this imaginary and design, engineers consider comfort as something stable and predefined, simply as data that the system must take into account but not as a social norm that could be negotiated and defined by and with users, changing over time and according to individual or collective experience.

Social dimensions of energy practices are well identified by sociologists and social anthropologists, particularly the role of social norms such as collectively stabilized definitions of what counts as clean or dirty, pure or impure within a group, a family, a society, or a country (Douglas 1966; Jack 2013), or what counts as a comfortable temperature (Shove 2003; Shove et al. 2014). Yet in the approach that smart building engineers take, comfort targets (i.e., setting the temperature at 20°C or 21°C) as well as climate targets (i.e., decreasing energy consumption and CO_2 emissions) go unquestioned. They are regarded as stable data, not as negotiable and dynamic. For instance, what about heating people rather than space by asking them to wear warm clothes or work in smaller rooms?

Energy efficiency means achieving a double objective: consuming less while maintaining the same level of comfort. This requires that users comply with the technical requirements of energy efficiency and interfere with it as little as possible (for example, not disturbing the mechanical and automated ventilation of the building by opening windows). Smart buildings convey an imaginary of a living organism parasitized by users, which creates a split vision of use. Users are either harmful parasites that threaten building efficiency, or they could become more beneficent parasites and help improve building performance.

Smart Users as Fallible Parameters of the Building

Smart buildings are governed by somewhat abstract promises and commitments made by global corporations or multilateral government bodies. While they are able to frame the problem and prioritize some solutions over others (Audétat et al. 2015; Borup et al. 2006; Brown and Michael 2003), these promises remain vague on technicalities.

According to engineers of Live TREE, they aim to do "what others have just been planning for years," such as Microsoft or IBM with their smart city projects: to test solutions in "real life" and "not in the lab." When the ideal becomes reality, when the lab experiment gets translated into a concrete construction site, the smart building becomes slightly messy: incompatible technical systems, incommensurable user needs, or unreliable sensors are just some of the problems encountered.

Whereas buildings are technically operational, engineers remain puzzled by how users seem reluctant to adopt the proper behaviours requested by the smart system they have designed. At best, the system presents a steep learning curve for users, who often need to adapt their individual practices and habits. For example, designers set the heating temperature and leave a narrow margin of adjustment, yet comfort varies with gender, age, weight, etc. Paradoxically, for some lead engineers on the project, a smart building is first and foremost meant to serve its users: it should be able to anticipate their needs with artificial intelligence algorithms, which means it should work better if they are not aware of its operations (Chamoin and Lenglet 2021). The building is thus expected to become smart enough – meaning fine-tuned and fed with enough data – to be able to think in place of the humans, or more accurately of an idealized human. It's not that inequalities and variety are actively negated by engineers, but, made imperceptible, they cannot be taken into account.

In such a strong deficit model, users are mostly passive and require training to behave as expected.[3] As STS scholars have shown in other fields such as cybersecurity, any malfunction of the system is then attributed to the supposed uneducated user as an easy way out of the complexity of the situation (Klimburg-Witjes and Wentland 2021, 1319). In this version of the user's "proper" role, engineers and designers secure their position as gatekeepers who set norms of practice and knowledge, inviting alternative takes and categories of actors the better to exclude them when it comes to technical decisions (Felt and Fochler 2010). In this perspective, the smart building is constantly at risk of becoming useless, as if the human component, which cannot be fine-tuned like other parameters, could degrade its proper functioning and ruin its architecture at any moment. Users are thus perceived as parasites who would like to benefit from the building when it suits their needs and deny its operations when they do not (Serres 1982).

---

3 Far from the desirability of "smart people" critically analysed by Mu in this volume.

The technological system is designed through a cost-benefit analysis made by the engineers to lower energy consumption.[4] Smart users of such an autonomous building, which exists in a fluctuating state of precarious instability and stabilization, are seen as fallible and uncontrollable parameters. They need to remain as passive as possible, adapting to the broader scheme as if human intelligence were threatening the smartness of the building by not being technically complacent enough. Here, the ideal user would ultimately be a user who isn't in the building.

Smart Users as Partners of the Building

Yet a different vision of smart users is inscribed in Live TREE and its smart building demonstrators. Although it is not necessarily less paternalistic than the version of things described in the previous section, it encompasses a range of tactics aiming at empowering users to actively participate in the program. To that end, rather than controlling their behaviours, engineers try to convince users of the merits of the project, help them appropriate its goals through pedagogical tools, and ultimately produce an informed smart user that can take ownership and build agency. That can mean decreasing their energy consumption when the building consumes more than it produces and delaying some tasks or energy-intensive needs until a moment when it would be more suitable for the building's efficiency.

Rather than creating space for users to contribute to the adjustment of technological systems and thus actively co-design them, as they had been invited to do during the early phases of the project in participative workshops (Wilkie and Michael 2009, 505), engineers distributed instructional materials. These included posters in office spaces explaining how to set the temperature (fig. 9.1) or screens that live-streamed energy consumption data in the form of charts and curves. Smartphone apps that send users notifications and alerts are also under consideration. Thus, when it comes to discussing users' engagement, Live TREE engineers have a wide repertoire for articulating building data with user participation: "provide information," "measure the compliance," "communicate on the project," "raise awareness about what is at stake," "get users involved," or "create the conditions for identification."

Nudging is another tool that gets increasingly (yet carefully) mobilized to inform, train, and influence users (Robyns et al. 2023). Typically

---

4 Marquis, in this volume, offers a compelling account of a similar push for smart farming in the name of sustainability.

Energy Transition to the Test of Parasitic Humans 101

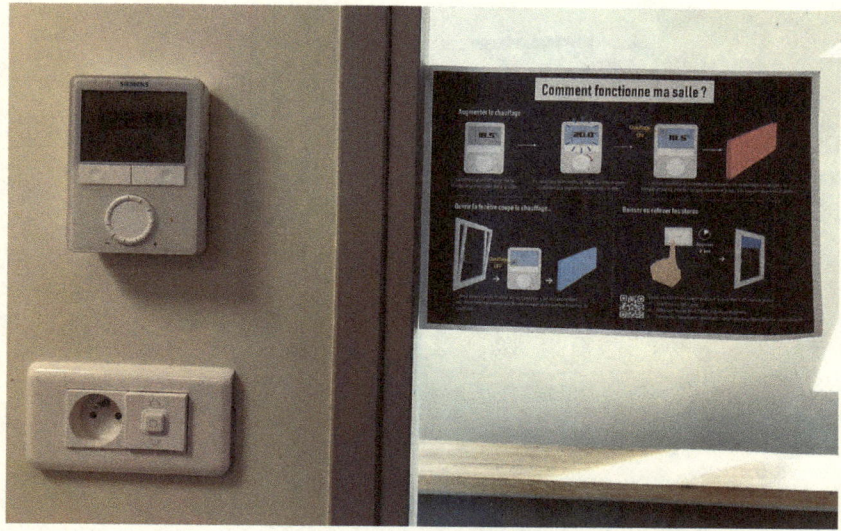

Figure 9.1. Posters in office spaces explaining how to set the temperature, open the windows, etc.

defined as a way of establishing an architecture of choice and pushing individuals to make advantageous decisions for their health or the environment without their feeling manipulated (Thaler and Sunstein 2008), nudging can take the form of tactful and joyful stickers strategically displayed above light switches. Figure 9.2 displays a sticker with a well-known French saying: "Turn off the light... This is not Versailles here!" Referring to the Royal Palace of Versailles, many French parents use this saying to admonish their children when they forget to turn off the lights or waste electricity.

There are lengthy discussions of topics such as governmentality, personal responsibility, rationalization, increased reflexivity, and modified architectures of choice in environmental studies (Dubuisson-Quellier 2016; Maniates 2001; Sahakian and Dobigny 2019). Nevertheless, according to Foucault (1979), governmentality is a dynamic process in which individuals are both the governed and the governing. Through the process of subjectivation, users can adopt the expected form of behaviour, resist, or get around it. Figure 9.3 shows one example of bypassing an expected behaviour by blocking the hydraulic mechanism designed to close doors and maintain temperature (reminiscent of Latour 1992). The photo was taken in the office of Live TREE lead engineers, where

Figure 9.2. An example of nudge: "Turn off the light… This is not Versailles here!"

Figure 9.3. "The Groom Is on Strike."

tinkering with the system and transgressing the very rules that they are trying to enforce was deemed necessary to fulfil the engineers' pedagogic goal of remaining open to questions and dialogue with users. Being a smart user ultimately touches upon crucial questions of distribution of expertise and power.

This suggests that the need for "smart users" is less a matter of shaping conduct to enact the desirable future promised by the smart building than one of empowering users so that they act as mediators and modulators of the engineer's expectations. For such a constantly evolving building, smart users need to demonstrate an ability to adapt, experiment, and improvise in order to find the right balance between performance, efficiency, and habitability. This suggests a more active attitude than the one articulated in the prior section. Here, the ideal or *smart* user would be the engineers themselves.

Conclusion: The Smartification of What?

Although not mutually exclusive nor equally committed, two visions of smart users are coproduced with two visions of the smart building. In the first case, a narrow view of smartness asks users to conform and adapt to building requirements as designed by engineers. Here, being "smart" means not hindering the proper functioning of smart technological systems. The ideal user is one that is simply not there. In the second case, "smartness" is better distributed among engineers and users, and the smart user is the one who knows how the building works as well as the engineer and adapts their behaviour and usages accordingly. But in both visions, usage, practices, habits, comfort, and social norms are not in question. Comfort (e.g., heating a room to 20°C) and energy efficiency are not up for debate. Can we feel comfortable at 18°C? How? By wearing warm clothes or drinking hot beverages? Is this feeling evenly distributed? Such smartification processes in a smart building illustrate an attempt to bring comfort and reach energy efficiency through universal technological solutions based on a standardized vision of users, their comfort, and their expectations.

Alternatively, smartness could be understood as a capacity to explore new paths, experiment with new practices, and invent new ways of doing. Sufficiency could be reached by collectively renegotiating social norms of comfort and by developing solutions and norms adapted to and fitting into a variety of contexts (e.g., climates, settings, age groups, and cultures). In this perspective, a smart user would be empowered and embedded in a collective effort of defining what consuming less involves, what is comfort, and what are the best ways to achieve sustainability. To facilitate this, engineers might benefit from ceasing to warily track down what users do wrong and thereby become more inclined to meet and listen to those who take the initiative, even if that means being caught by surprise.[5] The tensions between two contrasting understandings of smartification, as explored in this chapter, can help to sustain their

---

5 See also Sepehr, this volume.

generativity, provide resources to question what counts as smart and for whom, and prevent the transformation of buzzwords into watchwords.

References

Audétat, Marc, Gaïa Barazzetti, Gabriel Dorthe, Claude Joseph, Alain Kaufmann, and Dominique Vinck, eds. 2015. *Sciences et technologies émergentes: Pourquoi tant de promesses?* Hermann.

Bernard, H. Russell. 2018. *Research Methods in Anthropology: Qualitative and Quantitative Approaches.* 6th ed. Rowman & Littlefield.

Borup, Mads, Nik Brown, Kornelia Konrad, and Harro Van Lente. 2006. "The Sociology of Expectations in Science and Technology." *Technology Analysis & Strategic Management* 18 (3–4): 285–98. https://doi.org/10.1080/09537320600777002.

Brown, Nik, and Mike Michael. 2003. "A Sociology of Expectations: Retrospecting Prospects and Prospecting Retrospects." *Technology Analysis & Strategic Management* 15 (1): 3–18. https://doi.org/10.1080/0953732032000046024.

Chamoin, Julien, and Claude Lenglet. 2021. *Prospective, le podcast de Junia*, podcast, "Le bâtiment connecté est-il Intelligent?" December 17. https://podcast.ausha.co/prospective-le-podcast-de-junia/3-julien-chamoin-le-batiment-connecte-est-il-intelligent.

Cherry, C., C. Hopfe, B. MacGillivray, and N. Pidgeon. 2017. "Homes as Machines: Exploring Expert and Public Imaginaries of Low Carbon Housing Futures in the United Kingdom." *Energy Research & Social Science* 23 (January): 36–45. https://doi.org/10.1016/j.erss.2016.10.011.

Douglas, Mary. 1966. *Purity and Danger: An Analysis of Concepts of Pollution and Taboo.* Routledge and Kegan. https://doi.org/10.4324/9781315015811.

Dubey, Gérard. 2013. "Introduction." *Socio-Anthropologie*, no. 27, 9–20. https://doi.org/10.4000/socio-anthropologie.1403.

Dubuisson-Quellier, Sophie, ed. 2016. *Gouverner les conduites.* Presses de SciencesPo. https://doi.org/10.3917/scpo.dubui.2016.01.

Felt, Ulrike, and Maximilian Fochler. 2010. "Machineries for Making Publics: Inscribing and De-Scribing Publics in Public Engagement." *Minerva* 48 (3): 219–38. https://doi.org/10.1007/s11024-010-9155-x.

Foucault, Michel. 1979. *Discipline and Punish: The Birth of the Prison.* Vintage.

Garabuau-Moussaoui, Isabelle. 2014. "How People Work and Live in Energy Efficient Workplaces. Logics of Actions, Social Tensions and Real Issues for Occupants in Energy Efficient Office Buildings." *BEHAVE Conference Proceedings.* www.academia.edu/11078168/How_people_work_and_live_in_energy_efficient_workplaces_Logics_of_actions_social_tensions_and_real_issues_for_occupants_in_energy_efficient_office_buildings.

Gras, Alain. 2003. *Fragilité de la puissance. Se libérer de l'emprise technologique.* Fayard.

Houston, Lara, Jennifer Gabrys, and Helen Pritchard. 2019. "Breakdown in the Smart City: Exploring Workarounds with Urban-Sensing Practices and Technologies." *Science, Technology, & Human Values* 44 (5): 843–70. https://doi.org/10.1177/0162243919852677.

Jack, Tullia. 2013. "Nobody Was Dirty: Intervening in Inconspicuous Consumption of Laundry Routines." *Journal of Consumer Culture* 13 (3): 406–21. https://doi.org/10.1177/1469540513485272.

Jasanoff, Sheila, ed. 2004. *States of Knowledge: The Co-Production of Science and Social Order.* Routledge. https://doi.org/10.4324/9780203413845.

Jorgensen, Danny L. 2020. *Principles, Approaches and Issues in Participant Observation.* Routledge. https://doi.org/10.4324/9780367815080.

Klimburg-Witjes, Nina, and Alexander Wentland. 2021. "Hacking Humans? Social Engineering and the Construction of the 'Deficient User' in Cybersecurity Discourses." *Science, Technology, & Human Values* 46 (6): 1316–39. https://doi.org/10.1177/0162243921992844.

Latour, Bruno. 1992. "Where Are the Missing Masses? The Sociology of a Few Mundane Artifacts." In *Shaping Technology/Building Society: Studies in Sociotechnical Change,* edited by Wiebe E. Bijker and John Law. MIT Press.

Maniates, Michael F. 2001. "Individualization: Plant a Tree, Buy a Bike, Save the World?" *Global Environmental Politics* 1 (3): 31–52. https://doi.org/10.1162/152638001316881395.

Rifkin, Jeremy. 2011. *Third Industrial Revolution. How Lateral Power Is Transforming Energy, the Economy, and the World.* Palgrave Macmillan.

Robyns, Benoît, Claude Lenglet, Hervé Barry, and Malik Bozzo-Rey. 2023. *Smart Users for Energy and Societal Transition.* Wiley. https://doi.org/10.1002/9781394229895.

Sahakian, Marlyne, and Laure Dobigny. 2019. "From Governing Behaviour to Transformative Change: A Typology of Household Energy Initiatives in Switzerland." *Energy Policy* 129 (June): 1261–70. https://doi.org/10.1016/j.enpol.2019.03.027.

Serres, Michel. 1982. *The Parasite.* Translated by Lawrence R. Schehr. The Johns Hopkins University Press.

Shove, Elizabeth. 2003. *Comfort, Cleanliness and Convenience: The Social Organization of Normality.* Berg.

Shove, Elizabeth, Gordon Walker, and Sam Brown. 2014. "Material Culture, Room Temperature and the Social Organisation of Thermal Energy." *Journal of Material Culture* 19 (2): 113–24. https://doi.org/10.1177/1359183514525084.

Thaler, Richard H., and Cass R. Sunstein. 2008. *Nudge: Improving Decisions About Health, Wealth, and Happiness.* Yale University Press.

UNFCCC. 2015. *The Paris Agreement*. United Nations Framework Convention on Climate Change. https://unfccc.int/process-and-meetings/the-paris-agreement/the-paris-agreement.

Wilkie, Alex, and Mike Michael. 2009. "Expectation and Mobilisation: Enacting Future Users." *Science, Technology, & Human Values* 34 (4): 502–22. https://doi.org/10.1177/0162243908329188.

Winner, Langdon. 1986. *The Whale and the Reactor: A Search for Limits in an Age of High Technology*. University of Chicago Press.

10 Out of Thin Air: Vertiplaces

SIMON RABYNIUK

Introduction

Forms of mobility are constitutive of urbanism and shape how people inhabit cities (Söderström 2017). As such, the introduction of new modes of urban mobility enacts historic, often violent transformations of city form (Mumford 1961), produces new experiences and relationships within and of urban space (Simmel 1969), and extends the design of human settlements well beyond city limits (Hall 2014, 164–5). The stakes for urban mobility design are high as infrastructure distributes life-sustaining resources within a society (LaDuke and Cowen 2020).

Urban mobility remains essential for twenty-first-century urbanism and is a gateway issue for smart city initiatives (Halegoua 2020, 6). Increasingly, drones appear as a common trope in smart city imagery.[1] However, these visions of urban futurity fail to represent the broader, more ambitious digitalization of twenty-first-century civil aviation. Drone system builders strive to create automobile alternatives that leave the "congestion" of terrestrial streets behind (Corona 2021). Known as Advanced Air Mobility (AAM), which in the United States (US) is a national project fostering the creation of a commercial drone mobility market, this implicates wider social and spatial impacts for the whole of the built environment.

Vertiplaces are port infrastructure for drones. As interfaces between ground and sky, vertiplaces mediate the relationships between travellers, goods, and drone taxis or freight drones (Johnson 2021a).[2] Within the

---

1 For example, see Sidewalk Lab Toronto's images for Quayside, or architectural firm Carlo Ratti Associati's Anas Smart Road proposal, among others.
2 NASA and other AAM system builders commonly use the term electrical Vertical Takeoff and Landing Vehicle (eVTOL) to describe this broad category of next generation aviation vehicles. However, this chapter uses the term drone, drone taxis, or drone taxis and freight, instead reflecting a plainer description.

Figure 10.1. Vertiport Automation System (Vas) Airspace Ov-1 Diagram, 2020. Image Source: NASA.

US, there are no built examples of vertiplaces. However, the National Aeronautics and Space Administration's (NASA) Vertiport Series reveals its state of the art by discussing the autonomous performance of vertiplaces, as well as their form, scale, use cases, and integration into the built environment and regional networks. Differentiated by location, whether at an airport, urban centre, or suburb, vertiplaces support different amenities and traffic volumes (Daniels 2021). Like airports, they exist at the confluence of national US aviation regulations, state building codes, and municipal by-laws, in which architects and consultants interpret these standards for a specific site (Osborne 2022). Following the schematic design for these transportation nodes reveals the intentions of aerospace actors to shape the form and experience of near-horizon city-regions. The term city-region evokes a lineage of urban theory investigating how mobility, as one factor, participates in catalysing extended patterns of urbanization: linking city, country, and planet (Brenner 2014; Cronon 1992; Gottmann 1964; Lefebvre 2003). Since the First World War, nation-states have governed aviation; drones in the present are no different, except that their use of city skies forces a collision between the national and the local. Friction exists between AAM's conception in the abstract space of designing a national system and how that then transposes onto specific sites. AAM's planners negotiate this friction by plastering over – a process later described as a simplification – the social and spatial complexity of the city-region and its inhabitants.

This chapter focuses on the solid nature of architecture as an entry point for interpreting the design of AAM. It does so by tracing a series of five discussions encompassing the Vertiport Series convened by NASA's Community Integration Working Group (CIWG). This methodological choice follows a shift in mobility studies from attempting external, objective descriptions of transportation systems towards interpreting them from within (Söderström 2017, 198). As AAM is unrealized, interpreting it from within involves witnessing its planning process. Recorded verbal presentations, slides, and associated research reports, while publicly accessible, reveal insight into the closed and technocratic process used by NASA and the broader network of public and private actors in designing a new form of urban mobility.

The Architecture of Community Integration

In 2020, a set of four working groups convened by NASA began the next phase of introducing drone taxis and freight into the US. NASA's goal for AAM is no less than revolutionizing city and regional skies (NASA 2018). Succinctly named the Aircraft, Airspace, Community Integration, and Crosscutting Working Groups, each addresses current challenges for AAM through each group's unique lens. Broad themes include the automation of urban mobility, real-time situational awareness through ongoing sensing and data aggregation, the quantified management of risk, and discussions about the private ownership of critical infrastructure. These issues, as seen in establishing a commercial drone market, mirror smart city strategies, visions, and pitfalls.

A live poll during one working group session revealed the group's professional composition. Attendees included employees from the Federal Aviation Administration (FAA, the US civil aviation regulator), NASA, aerospace industry professionals, transportation planners, and state and municipal authorities (Johnson 2021b). NASA's representatives set the meeting's agenda and are responsible for inviting presenters. They also chair the meetings, which conclude by receiving questions from the specialist audience. The working groups do not actively complete research but instead serve as communication channels that frame questions and disseminate information, building capacity for AAM's industry stakeholders. This focus was affirmed during the series' introductory remarks when the facilitator acknowledged how topics could change to reflect whatever industry needed most immediately (Johnson 2021a).

With the launch of its Urban Air Mobility Grand Challenge, NASA made aspirational claims for a "safe, efficient, convenient, affordable, environmentally friendly, and accessible [AAM]" (NASA 2018). The Grand Challenge, while beyond the scope of this chapter, encompasses a series of industry demonstrations intended to prove AAM's viability. While announcing the rebranding of AAM, a NASA representative embellished these earlier claims, stating its goal as moving "people and cargo between places previously not served or underserved by aviation" (NASA 2020). These statements convey the sense that NASA envisions AAM as accessed by many, that it reduces transportation delays and the environmental effects of congestion. As such, the working groups focus on AAM's challenges in becoming an everyday form of urban mobility.

Tracing these issues demonstrates the entanglement between seemingly technical questions – how much noise propellers make or how drones share airspace, questions discussed in other working groups – and cities' social and lived dimensions. The CIWG most closely addresses urban space and a social perspective on AAM, albeit through standards and regulations, diagrams, and other representational tools that generalize and abstract urban space, sanitizing the social dimension. The Vertiplaces Series, as convened by the CIWG, confirms this view. Events are free to attend online, and their recordings are publicly available, in certain regards providing few barriers to access; however, during the first event, one presenter acknowledged his familiarity with many people in attendance, suggesting attendees had existing professional ties with AAM (Daniels 2021).

Series presenters discuss the FAA's process for creating vertiplace architectural standards (Bassey 2021). They also review automation studies that design vertiplaces for high-traffic volumes (Daniels 2021; Daniels et al. 2021). Additionally, they reflect on how vertiplace use models and vehicle design influence the creation of an open or monopolized AAM market (Alexander 2021; Corona 2021). The series also includes the presentation of a consumer survey investigating urban and rural residents' perceptions of possible vertiplace locations. The event's host identified how NASA's Aeronautics Research Mission Directorate – the institution overseeing AAM and facilitator of the Vertiport Series – did not typically perform social research and thus required collaboration with an external contractor (Nordstrom 2022). The final presentation shared schematic designs for vertiplaces that an architect created. Unlike earlier presentations that relied on diagrams for visualizing vertiplace operations, the architectural proposals rendered AAM's material, social, and spatial implications most clearly (Osborne 2022). Witnessing these presentations, one can interpret what NASA means by community integration. Implying a universalizing definition, NASA's use of community neither refers to a situated place nor a specific group of people. Instead, it evokes relationships between NASA, civil aviation regulators, and industry, with each party demonstrating an equal commitment to the commercialization of AAM.

Planning from Above

Legal historian Stuart Banner's (2008) study of the changing ownership and geographic boundaries of the twentieth-century sky observes aviation regulation lagging behind advancements in flight technology. In contrast, NASA's development of AAM takes a different approach by facilitating exchanges between regulators, researchers, and industry. This exchange fosters the integration of new vehicles, port infrastructures, concepts of air traffic management, and regulations for city skies. Transportation historian Christopher Wells (2013) documents similar forces entrenching the automobile in the mid-twentieth century. For AAM, citizens are absent except as potential consumers, reflecting the technocratic or top-down tradition of planning urban mobility that Wells documents for the automobile.

Political scientist James Scott's (1998) theories of legibility and simplification provide a useful way of interpreting grand-scale urban visions. Public and private actors participate in creating AAM, with NASA, itself publicly funded, having a central role. NASA's actions have an orienting effect that focuses different groups toward a vision of a commercial drone market that first serves the industry and then the public (Levitt et al. 2021, 6; NASA, n.d.). Scott uses the concept of legibility to theorize how states create representations of their populations as a precursor for governance. Statistical and geographic techniques such as censuses, cadastral maps, and security apparatuses describe populations in the present (Scott 1998, 343). For all of his attention to planning, Scott does not attend to how non-state actors participate in the success of grand state-led projects. Notably, the study of AAM reveals how the state also has practices for building confidence and a direction for a chosen future condition. Observing these practices, one can revise Scott's idea of legibility as projective legibility. NASA's actions of organizing the Urban Air Mobility Grand Challenge, publishing drone technology road maps, performing and commissioning AAM research, and organizing the working groups each create confidence and signal how private actors can contribute.

The second concept, simplification, studies how this planning happens and what its effects are. Simplification, Scott claims, is one reason why twentieth-century top-down planning projects failed. From his case studies, his findings emphasize how planners held limited conceptions of urban and rural spaces by ignoring the tacit, practical, or uncodified knowledge held by residents that made those places work (1998, 311). The concept of simplification reveals structural elisions in expert knowledge that, in Scott's view, make large-scale projects socially destructive if not accounted for. He observes state-led projects enacting irreversible transformations of the built and wider managed environments while falling short of their ambitious social or economic aims. Paralleling these planners' and designers' ambitions, NASA envisions equally distributing access to AAM (NASA 2018, 2020). Despite having egalitarian aspirations, Scott notes the consequences of these past masterplans' failures. As seen in the extreme examples of master-planned cities – historically, state capitals but today more often smart cities – failure emerges from creating inflexible spaces that do not reflect the complexity of human use. Instead, these projects manifest the erasure of context and the imposition of functional separations of uses (Scott 1998, 145).

Enacting Simplification

Planning practices enact simplification by focusing on monofunctional rather than ecological interpretations of space (Scott 1998, 346). An ecological view perceives multiple actors' uses and claims on and for space, revealing their relationality. In contrast, a monofunctional view associates one use for a space. Scott provides the example of viewing a forest as only lumber; however, for cities, single-use zoning delineating areas for living, working, commerce, and recreation provides another example. An ecological view of city skies may acknowledge property regimes like air rights, the translucency of residential towers, the seasonality of migratory birds, and urban microclimates with these legal, built, living, and atmospheric conditions creating an enmeshed urban assemblage (Farias and Blok 2016, 1). In contrast, NASA's monofunctional perspective interprets urban space through the capacities and affordances of the drone. Two ways NASA's planning accounts for claims on city skies are its acknowledgment of existing piloted forms of aviation and the need to control the height of properties adjacent to vertiplaces to preserve take-off and landing approaches (Alexander 2021). Stripping away spatial complexity renders city skies mostly empty and, therefore, available for commercial appropriation.

A monofunctional view of space instils a second form of simplification in master planning: designing for a standardized subject rather than people with shared and individual identities (Scott 1998, 346). As acknowledged by a NASA representative during the Vertiport Series, social research methods are not commonly used by the Aeronautics Research Mission Directorate (ARMD) (Nordstrom 2022). By ignoring questions about the inhabitation of urban space, future users are left undefined and, in some instances, merely as units in vertiplace throughput calculations. Having completed a consumer survey, ARMD has begun collecting urban and rural residents' perceptions of AAM. The survey first reveals a broad lack of awareness about AAM, which NASA puts forward as a near-horizon mobility choice. Second, AAM's greatest appeal is to high-income males (Nordstrom 2022). Do these findings challenge NASA's larger aim for an egalitarian AAM? The event presenter summarized the survey results without reflecting on how these findings could inform further work.

AAM planning introduces another dimension of simplification by focusing on the conceptual and technical challenges of introducing flight into cities rather than into its social life. By doing so, spatial proposals for AAM redesign the city-region, starting with the design of vertiplaces for the capacities and affordances of the drone. This makes urban space a by-product of vehicle performance. A historical view of railways or automobiles presents parallel examples of rails and roads transforming urban space in their image. Returning to historian Christopher Wells, he identifies how the US became a "car country," an automobile-centred landscape, by tracing the impact of 1950s infrastructure projects and the policy, incentives, and practices that spurred highway expansion (2013, xxxii). Two discussions during the Vertiplace Series similarly acknowledge the influence of AAM's next-generation vehicle's performance and standards on urban space.

In the first instance, a representative of the FAA stated how the regulator's architectural standards for vertiplaces would reflect the capacities and affordances of AAM vehicles (Bassey 2021). At the time of this presentation, over two hundred original equipment manufacturers (OEMs) – drone taxi and freight vehicle makers – were creating distinct models unique in their form, flight method, range, speed, carrying capacity, and method for refuelling. The FAA based its architectural standard on feedback about OEMs' vehicles. Initially, intellectual property claims limited response by the vehicle makers to the FAA. However, direct contact overcame this initial lack of a response. In mid-2022, the FAA released an interim advisory circle, a national standards document, providing future vertiplace builders with design guidance (FAA 2022). The document's first figure visualizes the "controlling dimension," a unit of measure generalizing the maximum scale imagined for drone taxis, which vertiplace dimensions are designed for (FAA 2024, 10). As a national standards document, it governs the traits of future US vertiplaces.

Vertiplace architectural standards are a by-product of vehicle form. Additionally, vehicle design presents implications for vertiplace use and its regional spread. A presentation by a director from an OEM reflected how vehicle differences impacted whether vertiplaces could have shared use by many drone taxi fleet operators – each with distinct vehicles – or singular use by one operator (Corona 2021). Identical to a taxi company, a fleet operator owns, operates, and maintains vehicles. The latter of these two conditions, wherein vertiplaces only serve one fleet operator, implies a more closed system. One scenario illustrates a fleet operator monopolizing a region by building vertiplaces tailored to their fleet of vehicles. A second scenario, in which competing fleet operators build their own vertiplace networks, implies a fractured urban landscape where each operator provides differential access across the region. A poll during the event recorded that most attendees – not all – favoured common access vertiplaces (Johnson 2021b). Designing a vertiplace for a single fleet operator presents a simplified task compared with creating one for common use by multiple operators (Corona 2021).

AAM geographically extends NASA's varied modes of simplification. Unlike most smart proposals, AAM's implications are broader than an urban precinct or city and imply the creation of smart regions, a smart nation, and transnational smart zones. These zones, outfitted with extensive sensor networks and control rooms, test flight automation. Often spanning several existing political boundaries, these zones extend civil aviation's authority, creating a new form of jurisdiction. Since 2018, the entire state of North Dakota has operated as a commercial drone test bed, demonstrating one example of smartness's expanding geographies (Kurtz 2020). In other US states, intercity corridors, campuses, towns, and border regions perform similar roles. While vertiplaces are not yet integrated into these zones, schematic planning for their architecture visualizes these extending networks.

The series features two discussions about a trade report that in parallel designed the automation of drone flights and vertiplace architecture. Automation is deemed essential for creating high-traffic vertiplaces (Daniels 2021; Daniels et al. 2021). One can infer that the report's authors imagine vertiplaces operating free from human intervention, as anticipated for the drones (Daniels et al. 2021). Taking a typological approach to the architecture, the report presents vertiplaces as a broad category and then defines vertihubs, vertiports, and vertistops as its subtypes. Each subtype's functionality and scale are aligned for common site conditions at airports, city centres, or suburban or ex-urban areas (Daniels et al. 2021). Graphically illustrated by multinational professional services firm Deloitte, a subcontractor on the report, each subtype is rendered as a diagrammatic isometric image. Visualized in the style of stock illustrations of the city, they generically present something speculative. These digital images communicate a simplified geographic concept for where to locate vertiplaces, their volume of traffic, and their amenities. Implicit here is how architecture serves as a vehicle for standardizing a set of values and effects across national geography.

Conclusion

NASA's emphasis on founding a commercial drone market introduces a final dimension of simplification: framing mobility as primarily a consumer service. From this singular focus, other forms of simplification then emerge, including the monofunctional rather than ecological view of space, design for an abstract standardized subject, and the city-region redesigned as a by-product of architectural and communication rules for the drone. Historically, introducing new forms of mobility has violently transformed urban space and people's experience of it, as well as expanding the geographic limits of urbanization. Historian Lewis Mumford viscerally communicates how nineteenth-century private railways ripped through Victorian city fabric (1961, 446). Georg Simmel's urban sociology studied the experience and effect on residents' mental lives from the introduction of omnibuses, railways, and streetcars (1969, 360). Town planners such as the early twentieth-century Regional Planning Association of America – counting Mumford as one member – proposed the dispersal of Manhattan's population across New York State, linking industrial satellite towns with early highway infrastructure (Hall 2014, 164–5). Rather than merely a consumer service, these examples demonstrate how designing a form of mobility creates a new social contract written through the concrete and code that formalizes urban space for the movement of people and things.

These examples also suggest re-engaging the complexity of urban space by acknowledging the related social and spatial impacts of new forms of mobility. By centring the creation of AAM within aerospace expertise and privileging types of collaborations focused on market creation, NASA leaves AAM's social and spatial realms simplified and thus, given Scott's work, vulnerable to failure. If NASA's aerospace expertise is misaligned with the challenges of designing for city-regions, then what role might spatial practitioners like architects, landscape architects, and urban designers contribute?

The final Vertiplace Series event presented one example of an architect translating vertiplace standards and diagrams into schematic architectural proposals (Osborne 2022). An important distinction between the two is that the architect interprets these diagrams for a specific client and site. Proposals for greenfield sites, rooftops, and a technology campus each choreograph the flows of passengers and vehicles. These projects are physically compact with the take-off and landing pads dominating each proposal. In one instance, the architect demonstrates regional awareness, diagramming connectivity between the technology campus and Atlantic City, Princeton, New Jersey, New York, and other smaller centres. As a service provider, the architect responds to the client's brief in each instance, designing vertiplaces for a resort and as a shuttle serving a tech campus.

NASA's emerging standards and regulations for vertiplaces participate in the *smartification of everything* in so far as they determine an operational logic for the envelope of space that encompasses US cities and that extends into surrounding regions. Again, returning to Wells, he observes that the main elements of the US's automobile-dominated landscape were set by 1956. In the more than half-century that has followed, maintaining and embellishing that system has dominated the US mobility agenda rather than rethinking car culture (2013, xxxii). This observation illustrates the temporality and inertia of urban mobility as a system, contrasting how one might experience it in the present against its longer-term formation. The Vertiplace Series reveals AAM's guiding interests, debates, and reports – one picture of its founding moments. Viewing AAM at finer resolution through its architectural diagrams and schematic proposals demonstrates that NASA's egalitarian AAM is blue-sky thinking.

References

Alexander, Rex. 2021. "Public vs. Private: A Use-Case Conundrum for AAM." NASA Advanced Air Mobility Community Integration Working Group, streamed July 1, 1 h., 29 min., Youtube. www.youtube.com/watch?v=m9gZxS0Wkws.

Banner, Stuart. 2008. *Who Owns the Sky? The Struggle to Control Airspace from the Wright Brothers On.* Harvard University Press. https://doi.org/10.4159/9780674020498.

Bassey, Robert. 2021. "FAA Vertiport Design Standards and Research Update." NASA Advanced Air Mobility Community Integration Working Group, streamed April 1, 1 h., 29 min., Youtube. www.youtube.com/watch?v=gefvBmGXcvo.

Brenner, Neil. 2014. *Implosions/Explosions: Towards a Study of Planetary Urbanization.* JOVIS. https://doi.org/10.1515/9783868598933.

Corona, Erick. 2021. "Private vs. Public Vertiports: An OEM's Perspective." NASA Advanced Air Mobility Community Integration Working Group, streamed July 1, 1 h., 29 min., Youtube. www.youtube.com/watch?v=m9gZxS0Wkws.

Cronon, William. 1992. *Nature's Metropolis: Chicago and the Great West.* W.W. Norton.

Daniels, Jonathan. 2021. "High-Density Vertiport Automation Study." NASA Advanced Air Mobility Community Integration Working Group, streamed April 1, 1 h., 29 min, Youtube. www.youtube.com/watch?v=gefvBmGXcvo.

Daniels, Jonathan, Sevan Mehrabian, Joseph Block, Stuart Wilson, Dwight DeCarme, and Bell Rueangvivatanakij. 2021. "Vertiport Automation System Architecture." NASA Advanced Air Mobility Community Integration Working Group, streamed September 2, 2 h., 4 min., Youtube. www.youtube.com/watch?v=hUcnCZDoeis.

FAA. 2022. "FAA Releases Vertiport Design Standards to Support the Safe Integration of Advanced Air Mobility Aircraft." September 26. www.faa.gov/newsroom/faa-releases-vertiport-design-standards-support-safe-integration-advanced-air-mobility.

— 2024. "Engineering Brief #105: Vertiport Design." https://www.faa.gov/airports/engineering/engineering_briefs/eb_105a_vertiports.

Farias, Ignacio, and Anders Blok, eds. 2016. *Urban Cosmopolitics: Agencements, Assemblies, Atmospheres.* Routledge. https://doi.org/10.4324/9781315748177.

Gottmann, Jean. 1964. *Megalopolis: The Urbanized North-Eastern Seaboard of the United States.* MIT Press. https://doi.org/10.7551/mitpress/4537.001.0001.

Halegoua, Germaine. 2020. *Smart Cities.* MIT Press. https://doi.org/10.7551/mitpress/11426.001.0001.

Hall, Peter. 2014. "The City in the Region: The Birth of Regional Planning: Edinburgh, New York, London, 1900–1940." In *Cities of Tomorrow*, 4th ed. Wiley Blackwell.

Johnson, Marcus. 2021a. "Vertiport Series Kickoff." NASA Advanced Air Mobility Community Integration Working Group, streamed April 1, 1 h., 29 min., Youtube. www.youtube.com/watch?v=gefvBmGXcvo.

— 2021b. "Public-Use vs. Private-Use Vertiport Facilities Overview." NASA Advanced Air Mobility Community Integration Working Group, streamed July 1, 1 h., 29 min., Youtube. www.youtube.com/watch?v=m9gZxS0Wkws.

Kurtz, Adam. 2020. "North Dakota Testbed Unveils Drone 'Interstate' System." *Grand Forks Herald*, November 2. www.govtech.com/fs/infrastructure/north-dakota-testbed-unveils-drone-interstate-system.html.

LaDuke, Winona, and Deborah Cowen. 2020. "Beyond Wiindigo Infrastructure." *South Atlantic Quarterly* 119, no. 2 (April): 243–68. https://doi.org/10.1215/00382876-8177747.

Lefebvre, Henri. 2003. *The Urban Revolution*. University of Minnesota Press.

Levitt, Ian, Nipa Phojanamongkolkij, Kevin Witzberger, Joseph Rios, and Annie Cheng. 2021. "UAM Airspace Research Roadmap." NASA. https://ntrs.nasa.gov/api/citations/20210019876/downloads/NASA-TM-20210019876Final.pdf.

Mumford, Lewis. 1961. *The City in History*. Harcourt.

NASA. 2018. "NASA Issues First Urban Air Mobility Grand Challenge Industry Request." www.unmannedairspace.info/uas-traffic-management-tenders/nasa-issues-first-urban-air-mobility-grand-challenge-industry-request/.

—. 2020. "One Word Change Expands NASA's Vision for Future Airspace Mobility." March 23. www.nasa.gov/aeroresearch/one-word-change-expands-nasas-vision-for-future-airspace.

—. n.d. "Advanced Air Mobility Mission Overview." https://www.nasa.gov/mission/aam/.

Nordstrom, Wyatt. 2022. "Optimal Locations for Air Mobility Vertiports." NASA Advanced Air Mobility Community Integration Working Group, streamed January 25, 1 h., 56 min. Youtube. www.youtube.com/watch?v=IITf_FOqrx4.

Osborne, Ted. 2022. "Vertiport Design." NASA Advanced Air Mobility Community Integration Working Group, streamed November 10, 1 h., 29 min., Youtube. www.youtube.com/watch?v=vmKZGqHoX38.

Scott, James C. 1998. *Seeing Like a State: How Certain Schemes to Improve the Human Condition Fail*. Yale University Press. https://doi.org/10.2307/j.ctvxkn7ds.

Simmel, Georg. 1969. "Sociology of the Senses." In *Introduction to the Science of Sociology*, edited by Robert E. Park and Ernest W. Burgess. The University of Chicago Press. https://www.gutenberg.org/files/28496/28496-h/28496-h.htm#Page_356.

Söderström, Ola. 2017. "Mobilities." In *Urban Theory: New Critical Perspectives*, edited by Mark Jayne and Kevin Ward. Routledge. https://doi.org/10.4324/9781315761206.

Wells, Christopher W. 2013. *Car Country: An Environmental History*. University of Washington Press. https://doi.org/10.1515/9780295804477.

11 What Is Optimized Farming? Exploring the Smartness Mandate in Canadian Agriculture

SARAH MARQUIS

Introduction

There is no doubt that the global food system, as it operates now, is unsustainable (Clapp and Scott 2018; Constance et al. 2018). It contributes to significant environmental problems including, but not limited to, greenhouse gas emissions, biodiversity loss, water and air pollution, and soil erosion. Responses to ecological crises in agriculture have been diverse; however, the concept of "smart farming" has emerged as a popular solution to problems in the food system (Streed et al. 2021). Smart farming is a nebulous, ever-evolving concept but usually refers to the increased use of digital technologies in agriculture, such as robotics, sensors, and a wide array of information and communications technologies, to create high-tech and data-intensive environments in the pursuit of increased productivity, profitability, and sustainability of agricultural systems (Miles 2019). The CEO of Precision AI, a Saskatchewan-based ag-tech start-up, expressed excitement around smart farming in the following way: "Artificial intelligence is probably the greatest game changer in food production in human history. The robots are going to be growing our food and that's not just aspirational geek talk" (Kirby 2022). Cracks in this facade of smart farming are becoming visible, however, as critical social scientists (Bronson and Knezevic 2016; Bronson 2022; Duncan et al. 2021; Stock and Gardezi 2021) are beginning to explore the ambiguous social consequences of the turn towards computational logics in agriculture.

Halpern and Mitchell's concept of the "smartness mandate" (2023) is especially useful in the critical analysis of smart farming as it demonstrates the logic of smartness according to its proponents as one that *shines* (much as Douglas-Jones describes in the foreword to this book) as a beacon beckoning us into a fantastical, high-tech world in which environmental problems will be dealt with by algorithms and automation.

The smartness mandate sets the stage for this chapter, which explores the drive towards optimization, of maximizing "good" outcomes and minimizing "bad" outcomes, and how this is shaping the politics and practices of agriculture. Furthermore, an analysis from Halpern's lens complicates the conclusion that "smart" agricultural systems are inherently and intrinsically sustainable. By exploring the ways in which the drive towards optimization is leveraged by those who are innovating and selling these emerging technologies, one can begin to identify gaps in the flows of logic that equate smart systems with sustainable ones.

This chapter will make the argument, informed by disciplines such as science and technology studies, that smartness and sustainability are distinct and separate processes, and the equating of these two concepts in smart farming industry discourse in Canada may have unintended social and environmental consequences. I will discuss the importance of the drive towards optimization in smart farming and how attempts to achieve this goal are being carried out, not to achieve environmental sustainability, though that is a stated goal, *but at its expense.* This argument is supported from my findings gathered through the analysis of corporate and sponsored content coming out of the smart farming industry in Canada, specifically from the Precision Agriculture Showcase, a semi-annual trade show hosted by the Canadian media company Farms.com, that is meant to inform farmers about new and emerging digital agriculture technologies. This evidence points to the fact that optimized agriculture does not beget sustainability; this is merely a discursive trick used by the proponents of smart farming to attract resources into the sector. The consequences could be dire in the context of a fraught global food system currently being impacted by catastrophic global climate change, with multiple food crises on the horizon (McGreevy et al. 2022).

The Smartness Mandate in Agriculture

"Smartification," according to its proponents, is a process facilitated by emergent computational and digital technologies producing a "more resilient human species – that is, a species able to absorb and survive environmental, economic, and security crises by means of perpetually optimising and adapting technologies" (Halpern et al. 2017, 107). Now, the turn towards "smartness" is visible almost everywhere, from smart cities to smart phones – smart devices and environments are ubiquitous. Smartness is identifiable through the presence of digital sensors that collect data, communication networks through which these technologies can communicate, and the technological capabilities to analyse and act upon the data through algorithms and automation (Sadowski 2020).

Smartification, the process of becoming "smart," is becoming particularly visible in Canadian agriculture, especially in response to environmental crises and threats to food insecurity caused by climate change, famine, inequitable food access, and geopolitical upheaval such as the war between Ukraine and Russia (McGreevy et al. 2022). Smart agriculture promises to resolve these overlapping and worsening fissures in a Canadian agricultural system that advertises itself to be one of the strongest in the world.

Halpern et al. provide a framework through which the logics of smartness can be mapped: "(1) The territory of smartness is the zone; (2) The (quasi-) agent of smartness is populations; (3) The key operation of smartness is optimization; (4) Smartness produces resilience" (2017, 112). One must look no further than the variable rate application (VRA) of inputs in crop agriculture to exemplify the ways in which this smartness mandate is being actively deployed in agriculture. Farmers can carry out VRA by using technologies to acquire higher levels of spatial data captured by remote and in-field sensors that can track the differences in productivity across different zones in a crop field. VRA then dictates that farmers precisely (this is why smart farming is also sometimes referred to as *precision agriculture*) tweak their management decisions based on the specific characteristics of each zone (Weersink et al. 2018). If we apply Halpern's framework to the example of VRA, then: (1) the territory of the smart farm is the management zone; (2) the (quasi-) agent of the smart farm is the crop population (Stock and Gardezi [2021], in fact, make the argument that both human and more-than-human populations, from farm managers to crops, pests, and weeds, are being managed by ag-tech firms); (3) the key operation of the smart farm is optimized levels of inputs, such as seeds, herbicides, or fertilizers: smart farming strategies are built around the idea that there are optimal amounts of these inputs to be used; and finally, (4) the smart farm produces resilience in the face of crises; VRA allows for ever-more efficient decision-making by farmers working within the context of global food insecurity and climate change. VRA exemplifies the smartification process in many ways, although it must be said that there are myriad technologies and techniques that are being deployed in the name of smartness. It is simply necessary, however, to look critically at the direction in which smartification is steering the food system as the outcome may not be as utopian as proponents of smart farming would like the world to believe (Daum 2021).

Before delving into the discourse of smart farming, however, it is necessary to understand the nuances and contested definitions of "sustainability" to understand how smart farming is leveraged in its name. Proponents for smart farming align their logic most plainly with "sustainable intensification," an idealized transformation of the food system in which enough food would be produced to feed a growing population and reductions in environmental impact would be prioritized (Miles 2019). According to certain scholars (see Godfray 2015), sustainable intensification squarely fits within productivist or neo-productivist categories of agricultural production. However, other conceptualizations of sustainable food systems exist, agroecology being one of the most notable. Agroecology prioritizes biodiversity by nurturing "holistic systems that create long-term fertility" (Constance 2018). There are many different manifestations of agroecological practices, including regenerative agriculture, permaculture, and organic farming. It is locally embedded within communities as opposed to being globalized; power is decentralized and distributed throughout local and regional actors (Horlings and Marsden 2011; Laforge et al. 2021). Agroecology, though not prioritized by national and global governance structures (Lajoie-O'Malley et al. 2020), is considered to be a necessary resistance to dominant forms of industrial agriculture (McGreevy et al. 2022). Moreover, sustainable agriculture in Canada cannot be truly conceptualized without an acknowledgment that settler colonialism and the theft of Indigenous land are foundational to the current system. Food studies scholars (Laforge et al. 2021) admit that agroecology in practice has yet to fully embody the values of Indigenous food sovereignty, however, the movement attempts to attend to questions of power that affect equity and justice in the food system.

Smart Farming Mythologies in Canada's Agriculture Technology Industry

In 2021, I attended (virtually, due to COVID-19) the Farms.com Precision Agriculture Showcase. When describing who should attend the conference, the website states: "Farmers who are interested in optimizing returns on inputs while preserving resources, reducing costs and reducing waste, all while improving yields and minimizing farming environmental impacts to the land" (Farms.com, n.d.). Other evidence for the drive towards the simultaneous optimization of agricultural profits, productivity, and sustainability is pervasive on ag-tech start-up websites, corporate press releases, and blogs. For example, Toronto-based start-up Ukko Agro has a simple mission statement: "Helping farmers grow more, sustainably" (2020). When it comes to the detailed exploration of what these technologies can do for farmers, sustainability is often framed as an obvious and inevitable outcome of farming smarter. Smartification is advertised, ultimately, in the name of increasing managerial control over the environment (Sadowski 2020). The apparent messaging from ag-tech start-ups is that smart control will enable farmers to simultaneously do away with concern for both the environment and productivity. Smartification will solve both problems.

An example of this discursive trend, that of equating "smart" and "sustainable," was illustrated in numerous presentations at the showcase. One presenter explained how smart technology, in this case, autonomous machinery, was inevitably going to lead to more sustainable futures: "So the only way we're going to grow higher yields, with less chemicals and less inputs is with technology. Precision application of that input, better management of the crop through sensor data, and then responding to that in crop management, better planting of the crop to make sure that it has maximum potential, all of this is only possible through technology." This presenter made it clear that smart technology is indeed necessary for lower-input agriculture that could feed a growing population, and that there is *no other way to do* sustainable agriculture. The preoccupation with optimization harkens back to Frederick Taylor's rational management, an ideological paradigm that began in the factory but was intentionally extended into agriculture, with the goal of optimizing productivity. Despite the presenter understanding smartification as an inevitable element of economic and environmental progress, we can also understand it as a continuation of industrialization in agriculture which was and continues to be a calculated political project (Fitzgerald 2003; Miles 2019).

Sometimes, the reference to sustainability was completely left out. It is in these cases that the logical fallacies of smart farming became apparent. For example, another presenter at the showcase began his presentation on soil mapping technology by referencing the definition of "precision agriculture" that is on the International Society of Precision Agriculture website: "Precision agriculture is a management strategy that takes account of temporal and spatial variability to improve sustainability of agriculture production." The presenter went on to paraphrase this definition: "That's looking at inputs, right, fertilizer ... seed, water, how can we grow a more productive crop with, let's be honest, more stresses on that production, right, climate, if it's more rain, less rain, more heat, less heat ... so how do we become more productive while looking at that spatial and, of course, temporal variability?" It is telling that the presenter removed the mention of sustainability from his interpretation of the definition of precision agriculture – the optimization of productivity takes precedence. Furthermore, it is notable that the website blurb for the 2022 Precision Agriculture Showcase explains that "the conference focus is not only on technology, but on the interpretation of data to implement future changes on the farm to maximize on-farm profits" (Farms.com). Again, all references to environmental sustainability have been replaced by a focus on profit maximization. These findings show that any reference to environmental sustainability is conceptualized, if at all, only as an efficient management of resources and inputs.

Equating sustainable and smart agriculture – and in some cases *replacing* sustainable with smart – is conceptually easy because the term sustainability, as discussed above, is so flexible, nebulous, and undefined (Constance et al. 2018; Scoones 2007). Sustainability can mean anything, so why can't it mean "smart farming" too? However, various scholars have called into question the promises of sustainability made by smart farming enthusiasts. For example, Streed et al. find that "even by their own aims, [smart farms] are unlikely to fundamentally alter the world's food system, or even contribute substantially to feeding people" (2021, 7). Bronson found that "digital agricultural tools are not regenerative but rather, at least in their current economic and legal infrastructures, are reproducing a number of social and cultural food system challenges" (2020, 336). My findings from the showcase illustrate these conclusions too; sustainable agriculture, if referenced at all by the ag-tech industry, is conceptualized simply as a minimization of inputs. There is little reference to the structural problems of agriculture that, for example, agroecologists are attempting to contend with. Additionally, many critical social science scholars (see also Klerkx et al. 2019) find that the promises that smart farming makes, especially with relation to sustainability, contain untenable logical fallacies. Furthermore, key thinkers in environmental sustainability see beyond simply efficiency as the main function of sustainability. Amartya Sen, for example, contributes to the concept of sustainability put forth in the infamous Brundtland Commission Report by contending that "a fuller concept of sustainability has to aim at sustaining human freedoms, rather than only at our ability to fulfill our felt needs" (2013). It should be questioned how substantively smart farming can sustain "human freedoms." In terms of sustainable agriculture, Vandana and Kartikey Shiva critique the new wave of smartification as a paradigm that prioritizes control and surveillance of the environment (and the farmer) while advocating for a regenerative relationship with the land, one that does not see nature and humans as separate, but interconnected and relational (Shiva and Shiva 2020).

Halpern explains that the drive towards optimization, crucial in the process of smartification, creates the "fantasy of stretching finite resources [such as land and water] to infinite horizons through big data and artificial intelligences" (Halpern 2021, 248). Here I argue that smart farming is being taken up within the space of industrial agriculture because it helps to construct the fantasy that business-as-usual agriculture can continue without harm; this myth appears more attractive than more radical transformations that could be brought about through agroecology. By equating, and in some cases, effectively replacing the concept of "sustainable agriculture" with "optimized agriculture," proponents of smart farming can refocus their audience's attention on specific digital technologies instead of the potential need to make structural changes in the foundations of the agricultural system (Clapp and Scott 2018). It is important, however, to consider the implications that these imaginaries have in policymaking spaces. For example, the Canadian government, in their 2022 budget, is pouring financial resources into the ag-tech space with the reasoning that "[precision agriculture] technologies can help reduce [carbon] emissions and save farmers both time and money" (Department of Finance 2022). The International Panel on Climate Change has also advocated for the optimization and scaling up of agricultural practices using technological innovations (Stock and Gardezi 2021). One must consider the ways in which flows of private *and* public capital into the smart farming space have material and financial implications for farmers who do not subscribe to the belief that their farming practice needs to be "smart."

McKelvey and Neves explore the social practices and harms that are produced by the emerging desire for optimality in all facets of society. The optimization process (because it is a process, not an outcome – a system never, in actuality, reaches the optimum) relies on "longstanding colonial and scientific knowledges that apprehend self and social determination through the lens of development, progress, innovation and perfection" (2021, 97). Considering the racist, anti-Indigenous legacy of Canadian agriculture (Laforge et al. 2021; Rotz 2017), the value system exemplified and solidified by smart farming is of special importance. The need to control, dominate, own, and manage the natural environment is foundational to the history of Canadian agriculture (Rotz 2017), which is, in turn, foundational to the turn towards smart farming. In fact, the technologies that define smart farming environments are those that help the farmer "control the growing environment" (Streed et al. 2021). With VRA technologies, for example, there is no consideration that another way to manage crops might be to do away with harmful inputs altogether. As McKelvey and Neves state: "optimization is a colonial desire to control *with* and *through* technology" (2021, 103; emphasis in original). Smart farming seems to be locking in the settler colonial approach to agriculture as environmental control. In doing so, it locks out alternative forms of agriculture, such as agroecology. Not only is smart farming biased towards larger farms simply in the material realities of the technologies that are marketed towards farmers (Miles 2019), but it locks out values such as "community, connection to the land, awareness of ecological relationships, and the distinctiveness of regional foods" (Streed et al. 2021). The fact that these values are mainstays of agroecology and Indigenous relationships to food systems is no accident.

The Consequences of "Computation as Salvation"

Halpern discusses the ways in which there is a present tendency, in moments of crisis, to "turn to computation as salvation" (2021, 228). She talks about how the drive to optimize any resource extraction activity is defined by the idea that a "finitude of resources can be addressed through an infinity of data" (242). Proponents of smart farming operate by this logic, attempting to situate it as humanity's salvation in the face of diverse crises. As Mu articulates in an earlier contribution to this section: "the 'smart city' is a practice to hedge uncertain futures" (59), I argue that so, too, might be the "smart farm"; the message is that if only farming were more optimized, more streamlined, more efficient, then the food system could be made stable, resilient, and harmless. However, I argue that this cannot be true, as smart farming, as it is characterized by the ag-tech industry, does not address the structural unsustainability of an agricultural system that is firmly embedded in colonial capitalism (Laforge et al. 2021). Alternatively, agroecology entails holistic farm management as opposed to breaking the environment down into management zones with the goal of controlling its variables. Furthermore, the agroecological movement attempts to resist hegemonic power in the food system and decentralize this power to local actors (Horlings and Marsden 2011). It will be important to continue to foster emerging research into how technologies that are currently being used in service of smart farming could be reconstituted in agroecological environments (Bronson 2020; Duff et al. 2021). But it will also be important to watch how smart farming evolves and ask: What worlds are being created by smart farming? Are these worlds equitable? Could they be sustainable? Most importantly, who is smart farming optimal for? We should, perhaps, be imagining food systems that are not "smart," but that are founded on care, regeneration, community well-being and ecological diversity (McGreevy et al. 2022).

References

Bronson, Kelly. 2020. "A Digital 'Revolution' in Agriculture?" In *Routledge Handbook of Sustainable and Regenerative Food Systems*, edited by Jessica Duncan, Michael Carolan, and Johannes S.C. Wiskerke, 1st ed. Routledge. https://doi.org/10:4324/9780429466823-24.

— 2022. *The Immaculate Conception of Data: Agribusiness, Activists, and Their Shared Politics of the Future.* McGill-Queen's University Press. https://doi.org/10.1515/9780228012535.

Bronson, Kelly, and Irena Knezevic. 2016. "Big Data in Food and Agriculture." *Big Data and Society* 3 (1): 1–5. https://doi.org/10.1177/2053951716648174.

Clapp, Jennifer, and Caitlin Scott. 2018. "The Global Environmental Politics of Food." *Global Environmental Politics* 18 (2): 1–11. https://doi.org/10.1162/glep_a_00464.

Constance, Douglas, Jason Konefal, and Maki Hatanaka, eds. 2018. *Contested Sustainability Discourses in the Agrifood System.* Earthscan Food and Agriculture Series. Routledge, Taylor and Francis Group. https://doi.org/10.4324/9781315161297.

Daum, Thomas. 2021. "Farm Robots: Ecological Utopia or Dystopia?" *Trends in Ecology & Evolution* 36 (9): 774–7. https://doi.org/10.1016/j.tree.2021.06.002.

Department of Finance, Government of Canada. 2022. *Budget 2022: A Plan to Grow Our Economy and Make Life More Affordable.* https://www.budget.canada.ca/2022/home-accueil-en.html.

Duff, Hannah, Paul B. Hegedus, Sasha Loewen, Thomas Bass, and Bruce D. Maxwell. 2021. "Precision Agroecology." *Sustainability* 14 (1): 106. https://doi.org/10.3390/su14010106.

Duncan, Emily, Alesandros Glaros, Dennis Z. Ross, and Eric Nost. 2021. "New but for Whom? Discourses of Innovation in Precision Agriculture." *Agriculture and Human Values* 38 (4): 1181–99. https://doi.org/10.1007/s10460-021-10244-8.

Farms.com. n.d. "2021 Virtual Precision Agriculture Conference & AG Technology Showcase." Accessed August 15, 2022. www.farms.com/precision-agriculture/conferences/virtual-precision-ag-conference-2021/.

Fitzgerald, Deborah. 2003. *Every Farm a Factory: The Industrial Ideal in American Agriculture.* Yale University Press. https://doi.org/10.12987/yale/9780300088137.001.0001.

Godfray, H. Charles. 2015. "The Debate over Sustainable Intensification." *Food Security* 7 (2): 199–208. https://doi.org/10.1007/s12571-015-0424-2.

Halpern, Orit. 2021. "Planetary Intelligence." In *The Cultural Life of Machine Learning: An Incursion into Critical AI Studies*, edited by Jonathan Roberge and Michael Castelle. Palgrave Macmillan. https://doi.org/10.1007/978-3-030-56286-1_8.

Halpern, Orit, and Robert J. Mitchell. 2023. *The Smartness Mandate.* The MIT Press. https://doi.org/10.7551/mitpress/14623.001.0001.

Halpern, Orit, Robert J. Mitchell, and Bernard Dionysius Geoghegan. 2017. "The Smartness Mandate: Notes toward a Critique." *Grey Room*, no. 68 (Summer): 106–29. https://doi.org/10.1162/GREY_a_00221.

Horlings, L.G., and T.K. Marsden. 2011. "Towards the Real Green Revolution? Exploring the Conceptual Dimensions of a New Ecological Modernisation of Agriculture That Could 'Feed the World.'" *Global Environmental Change* 21 (2): 441–52. https://doi.org/10.1016/j.gloenvcha.2011.01.004.

Kirby, Jason. 2022. "The Agricultural Revolution Is Here: Will Canada Keep Up and Invest in Agtech?" *The Globe and Mail.* August 5. www.theglobeandmail.com/business/article-as-agtech-surges-can-canada-close-the-gap/.

Klerkx, Laurens, Emma Jakku, and Pierre Labarthe. 2019. "A Review of Social Science on Digital Agriculture, Smart Farming and Agriculture 4.0: New Contributions and a Future Research Agenda." *NJAS – Wageningen Journal of Life Sciences* 90–1 (1): 1–16. https://doi.org/10.1016/j.njas.2019.100315.

Laforge, Julia M.L., Bryan Dale, Charles Z. Levkoe, and Faris Ahmed. 2021. "The Future of Agroecology in Canada: Embracing the Politics of Food Sovereignty." *Journal of Rural Studies* 81:194–202. https://doi.org/10.1016/j.jrurstud.2020.10.025.

Lajoie-O'Malley, Alana, Kelly Bronson, Simone van der Burg, and Laurens Klerkx. 2020. "The Future(s) of Digital Agriculture and Sustainable Food Systems: An Analysis of High-Level Policy Documents." *Ecosystem Services* 45:1–12. https://doi.org/10.1016/j.ecoser.2020.101183.

McGreevy, Steven R., Christoph D.D. Rupprecht, Daniel Niles, Arnim Wiek, Michael Carolan, Giorgos Kallis, Kanang Kantamaturapoj, A. Mangnus, P. Jehlička, O. Taherzadeh, and M. Sahakian. 2022. "Sustainable Agrifood Systems for a Post-Growth World." *Nature Sustainability* 5 (12): 1011–17. https://doi.org/10.1038/s41893-022-00933-5.

McKelvey, Fenwick, and Joshua Neves. 2021. "Introduction: Optimization and Its Discontents." *Review of Communication* 21 (2): 95–112. https://doi.org/10.1080/15358593.2021.1936143.

Miles, Christopher. 2019. "The Combine Will Tell the Truth: On Precision Agriculture and Algorithmic Rationality." *Big Data & Society* 6 (1): 1–12. https://doi.org/10.1177/2053951719849444.

Rotz, Sarah. 2017. "'They Took Our Beads, It Was a Fair Trade, Get over It': Settler Colonial Logics, Racial Hierarchies and Material Dominance in Canadian Agriculture." *Geoforum* 82 (June): 158–69. https://doi.org/10.1016/j.geoforum.2017.04.010.

Sadowski, Jathan. 2020. *Too Smart: How Digital Capitalism is Extracting Data, Controlling Our Lives, and Taking Over the World.* MIT Press. https://doi.org/10.7551/mitpress/12240.001.0001.

Scoones, Ian. 2007. "Sustainability." *Development in Practice* 17 (4–5): 589–96. https://doi.org/10.1080/09614520701469609.

Sen, Amartya. 2013. "The Ends and Means of Sustainability." *Journal of Human Development and Capabilities* 14 (1): 6–20. https://doi.org/10.1080/19452829.2012.747492.

Shiva, Vandana, and Kartikey Shiva. 2020. *Oneness vs. the 1%: Shattering Illusions, Seeding Freedom.* Chelsea Green.

Stock, Ryan, and Maaz Gardezi. 2021. "Make Bloom and Let Wither: Biopolitics of Precision Agriculture at the Dawn of Surveillance Capitalism." *Geoforum* 122 (June): 193–203. https://doi.org/10.1016/j.geoforum.2021.04.014.

Streed, Adam, Michael Kantar, Bill Tomlinson, and Barath Raghavan. 2021. "How Sustainable Is the Smart Farm?" *LIMITS Workshop on Computing within Limits.* https://doi.org/10.21428/bf6fb269.f2d0adaf.

Ukko Agro. n.d. "Ukko Agro: Bringing Actionable Insights to Your Digitized Acres." https://ukko.ag/.

Weersink, Alfons, Evan Fraser, David Pannell, Emily Duncan, and Sarah Rotz. 2018. "Opportunities and Challenges for Big Data in Agricultural and Environmental Analysis." *Annual Review of Resource Economics* 10 (1): 19–37. https://doi.org/10.1146/annurev-resource-100516-053654.

PART THREE

Smartification as Boundary Work

12 Introduction to Smartification as Boundary Work

MASCHA GUGGANIG

"Smartification as Boundary Work" takes up a classic STS concept to highlight the multifarious ways in which actors draw and contest boundaries around smartness. As Thomas Gieryn asserted three decades ago in the case of science:

> Pragmatic demarcations of science from non-science are driven by a social interest in claiming, expanding, protecting, monopolizing, usurping, denying, or restricting the cognitive authority of science. (Gieryn 1995, 405)

Said differently, "science" does not have an inherent objectivity that exists "out there" to be discovered but is socially and culturally situated and constructed, requiring the constant act of boundary drawing – often by scientists – to demarcate it as space of epistemic authority from other forms of knowledges, questions, and people (Gieryn 1995; see also Harambam and Aupers 2015). What Gieryn describes for science is equally pertinent for smartness: it is constructed and situated through inclusion and exclusion, manifesting in how different actors define, defend, and contest technoscientific smartness of spaces, objects, and people (see Klimburg-Witjes et al. 2021). Consequently, they render other actors ignorant, idiotic, or "dumb" (Halpern and Mitchell 2023; Tironi and Valderrama 2018) or more generally demarcate them from forms of smartness that are entwined with human and more-than-human multisensorial awareness. Consequently, boundary drawing around smartness can be studied as a smartification practice and process.

When studying the boundaries drawn around smartness, a sensibility for spatiality is helpful. Critical scholars and artists of smartification can offer useful analytical tools to trace how certain actors and processes aim to both disguise and highlight such boundaries to keep certain places (and their inhabitants) in and out. In chapter 13, architecture scholar Ali Fard starts us off with troubling the very notion of drawing boundaries around the smart city. While key actors entertain a constant boundary drawing around the smart city as a manageable "archipelago in a sea of an undefined periphery" (139), his analysis shows that smartification processes are "unbounded" as the locations of necessary material for its palimpsest, "the cloud," are far and wide: data centres in rural and peri-urban areas, extraction sites for rare earth minerals, and dumping grounds far from smart city dwellers. Such critical analysis highlights the porosity and dependability of smartness (and cities) while pointing to the continuous need to analyse the global material geographies that upend any illusion of boundedness (see also Unknown Fields Division 2015). The next contribution is an artistic research, where STS scholar Sebastian Bornschlegl discusses his Visual Vignette of a smart city project in Vienna, Austria. Using this novel format, which allows for a creative recombination of word and image (Gugganig and Douglas-Jones 2021), Bornschlegl collages various, partially connected oases of smartness within the smart city of Aspern. The open-ended, artistic nature of the Visual Vignette allows the author to expand upon the more common format of a photo essay in two ways: first, the collage-like recombination of image and text reflects the fractured, at times digitally disconnected nature of a smart city to visualize – and critique – it as anything but a smooth operation. Second, the methodical transgression of the boundary of image and text likewise makes us pause to question how "smartness" operates on a visual, sensorial level.

In chapter 15, anthropologist Hannah Carlan traces how a smart city scheme in Dharamshala, India, comes to be by drawing boundaries. While most proponents of smart cities define said space by fencing off undesirable areas, people, and animals, Carlan shows how such undesired subjects and spaces are rather extracted from within the physical boundaries of the smart city. Dogs trapped in "smart" dustbins where they often end up dying stand as a tragic metaphor for being trapped and subsequently "digested" by the logic and means of smartness and overall processes of smartification. This rendering of certain entities as disposable is anything but a by-product of smartness; as Carlan argues, it is its very raison d'être. Part 3 ends with the reprint of the zine *Wal*Smartification: Considering the Superimposition of Dockless Shared Electric Scooters* on Fayetteville by poet Devin Shepherd and artist Juliette Walker. Through its evocative risograph print, the artists display and provoke us to consider smart scooters as urban clutter and thus as a tale of neoliberal "Walsmarting" or "free-market smartness" that is degrading public space in Arkansas, US. Through their visually enticing juxtaposition of blue and yellow scooters and sceneries in the form of artefacted duotone images, they beautifully highlight the unfittingness of scooters in public urban spaces. Similar to Kim's analysis of computistic ideas of deletion, formatting, and uploading that manifest in clearing land (chapter 6), Shepherd and Walker highlight the boundaries in the copy-and-paste logic of introducing scooters in the city. The underlying undemocratic logic – being able to use e-scooters by deciphering the e-scooter system and signs, having access to a cell phone – also draws boundaries around digitally literate "smart" citizens and spaces, consequently allowing the cluttering of public spaces.

References

Gieryn, Thomas F. 1995. "Boundaries of Science." In *Handbook of Science and Technology Studies*, edited by Sheila Jasanoff, Gerald E. Markle, James C. Petersen, and Trevor Pinch. Sage.

Gugganig, Mascha, and Rachel Douglas-Jones. 2021. "Visual Vignettes." In *Sensing In/Security: Sensors as Transnational Security Infrastructures*, edited by N. Klimburg-Witjes, N. Poechhacker, and G.C. Bowker. Mattering Press.

Halpern, Orit, and Robert Mitchell. 2023. *The Smartness Mandate*. MIT Press. https://doi.org/10.7551/mitpress/14623.001.0001.

Harambam, Jaron, and Stef Aupers. 2015. "Contesting Epistemic Authority: Conspiracy Theories on the Boundaries of Science." *Public Understanding of Science* 24 (4): 466–80. https://doi.org/10.1177/0963662514559891.

Klimburg-Witjes, Nina, Nikolaus Poechhacker, and Geoffrey C. Bowker, eds. 2021. *Sensing In/Security: Sensors as Transnational Security Infrastructures*. Mattering Press. https://doi.org/10.28938/9781912729111.

Tironi, Martín, and Matías Valderrama. 2018. "Unpacking a Citizen Self-Tracking Device: Smartness and Idiocy in the Accumulation of Cycling Mobility Data." *Environment and Planning D: Society and Space* 36 (2): 294–312. https://doi.org/10.1177/0263775817744781.

Unknown Fields Division. 2015. "Rare Earthenware." Film, Vimeo, 7 min. https://vimeo.com/124621603.

13 Reframing Smart Urbanism

ALI FARD

The technologically mediated transformations that are collectively referred to as "smart" are increasingly dictating the design and management of urban environments. But in discussing the influence of technology on the built environment within architecture and urbanism, a deeper engagement with the geographies that materially support these technologies seems glaringly missing. Instead, a sense of ambiguity permeates the conceptual core of the smart city and limits its perception to only highly visible projects within urban cores. Even within these highly concentrated moments, "smartness" applies to a whole host of situations and at various scales. Smartness could relate to the infiltration and monitoring of elements of architecture and the city with sensors and digital technology; it could describe how urban technical systems are made more efficient through networked technology; it could allude to the technologically mediated processes of urban governance and urban management; or, as with holistic "smartification" campaigns, a combination of all of the above. Then, how do we, as urban scholars and practitioners, delineate the smart city? If vagueness permeates the conceptual base of smart urbanism, what material realities and operational geographies help frame the sites and processes of smart cities?

Even with the ambiguity surrounding smart cities, two perspectives tend to dominate the perception of "smartness" in architecture and urbanism. On one side, the "smartification" of the city through technology is seen as an external force and an existential threat to architecture, which has traditionally been associated with the design and the visioning of the built environment. Rem Koolhaas, one of contemporary architecture's most prominent voices, sees architecture's current entanglement with digital technologies as "the most radical change within the discipline since the confluence of modernism and industrial production in the early twentieth century" (Koolhass 2015). As someone with a penchant for manifestos, Koolhaas believes that the "smart" city has stepped into the vacuum created by architecture's diminishing agency within global urbanization processes. For him, the "gradual colonisation of architecture" through a "stealthy infiltration of [its] constituent elements" amounts to a fundamental transference of authority over the organization of the built environment from architects and designers towards private technology firms and capitalist interests (Koolhass 2015).

On the other side, others, like the historian of architecture and engineering Antoine Picon, believe that the "smart" city as a conceptual framework cannot be reduced to a technological takeover of cities. Sifting through several critical accounts of smart urbanism, Picon defines the notion of the "smart" city as both an ideal and a process (Picon 2015). As an ideal, the smart city is one in which the proliferation of digital tools entails the optimization of the efficiency, functioning, and sustainability of the city and enables a certain quality of life for its inhabitants. The smart city is also a process through which urban environments become more intelligent through the influence of several economic, technological, and social transformations and the actions of human and non-human actors who are increasingly entangled within complex technological interrelationships. For Picon, this emerging understanding of the city touches on a more fundamental shift in the conception of cities. The advent of urban intelligence is leading to a transition from the networked city to the event city. In this conception, events and occurrences play a more significant role in urban experiences and the management of cities (Picon 2015, 51).

Both of these perspectives, however, focus solely on the after-effect of "smartification" as a process that is transforming the practices and processes of urbanization within cities. What if we flip the question? What if instead of asking how technological change is transforming the city, we ask how technology corporations – as dominant agents within the smart city – are positioning themselves as harbingers of new transformative forms and processes of urbanization? What if we ask how "smartness" is produced? What if we follow data – the lifeblood of smart urbanism?

Data and Territorial Hybridity of Platforms

Data and its continual regeneration are essential to the operations of the smart city. As mediators between the contemporary information society and its data, technology platforms play a massive role in the expansion of a global material geography that supports these data processes as well as the infrastructural support systems that the smart city is built upon. Platforms are sociotechnical constructions. They are co-produced through the expansion of technical infrastructure necessary for contemporary data regimes, as well as sociocultural positioning of themselves and their products at the forefront of discussions of technology and social progress (Jasanoff 2004, 2015). The construction of data platforms entails a particular territorial hybridity, which puts data-driven control and management of concentrated moments of urbanization in cities, in direct relationship with infrastructural extension and increasing operationalization of the erstwhile non-urban periphery.

On the one hand, platforms treat the city as a frontier of data-driven projects that attempt to deal with a variety of issues facing contemporary urbanization. From parking spaces and health care to climate change and democracy, nothing seems to be outside the purview of tech. In turn, marketing campaigns actively re-present the smart city's spatial projects as part of a singular, clean, objective, efficient, and transformative platform logic. The urban projects of tech platforms are portrayed as highly defined entities with clear boundaries that operate on the city as a wholly knowable, manageable, and ultimately controllable archipelago in a sea of undefined periphery. This further imbues the urban operations of data platforms with techno-utopian ideologies of progress that attempt to justify the colonization of everyday life through technology as a necessary consequence of responding to the socio-environmental challenges of urbanization (Greenfield 2017).

On the other hand, the takeover of cities by technology platforms is predicated on the buildup of computing resources and the enclosure of material geographies that are often situated well outside of cities. The ease-of-use and the "fluidity" by which many smart projects are characterized continually materialize in the very physical data infrastructures that enable the daily operations of platforms. In other words, the operational fluidity of the smart city is grounded on the "spatial fixity" that is at the core of capitalism's tendency towards geographic and territorial expansion (Harvey 2001). These are the infrastructural spaces that mediate the urban operations of platforms and their data supremacy. This is a global operational landscape dominated by cables, data centres, extraction sites, tech campuses, manufacturing plants, and dumping sites that literally tether the cloud to the ground.

However, the ambiguity and the abstraction that surrounds many smart projects effectively obscures the means and relations of their production as well as the power dynamics that are instituted and maintained through them – even to the point that we forget that every click, every sensor, every screen, and every AI platform within the smart city is ultimately linked up to clouds of data that depend on infrastructural landscapes and material geographies with inherent environmental intricacies and sociopolitical complexities. As much as the city and its organization figure within smart city discussions, the material necessities and infrastructural capacities of the smart city extend well beyond its boundaries. In fact, the operations of the smart city are dependent on the spatial (re)production of planetary-scale landscapes of extraction, storage, and processing, as well as infrastructural geographies of circulation that are necessary for the continual expansion of data platforms.

The production of this data geography is an intrinsically material process, from the embedded sensors that collect urban data and the servers that store it, to the cables that transmit data and the electronic devices that present it. As the processes at the core of platform capitalism demand continual territorial expansion, the competition over the increasingly sparse resources that materially support data economies will entail an ever-expanding operational landscape (Srnicek 2017). Google, for example, has spent more than $30 billion on its global cloud infrastructure.[1] Amazon has spent $35 billion on data centres in Northern Virginia alone (Miller 2021). Microsoft, Amazon, Google, and Meta have invested so heavily in undersea fibre-optic lines that their private networks' bandwidth capacity now completely overshadows the backbone of the internet (Mauldin 2016). Most, if not all, of these developments are taking place outside of traditional urban areas, that is, cities. Yet it is these very same operational geographies that tech platforms leverage to exert their considerable computational power in cities.

In this context, where do we locate the smart city? In the concentrated population centres where showpiece projects and test-bed urbanism propagate a future that is not very different from our present, only more technologically decked out (Halpern et al. 2013)? Or do we look to the lithium mines of Chile, the fibre-optic cables that line the Atlantic, or the data farms that dot the struggling industrial towns of middle America (Starosielski 2015b; Burrell 2020), the places where "smartness" is materially produced?

Remapping "Smartness"

To fully capture the urban impact of technology platforms, we need to remap the footprints of "smartness." This emerging map goes well beyond the city to situate the technical positivism of smart cities within

---

[1] Eric Schmidt, the Chairman of Alphabet, mentioned the $30 billion investment in Google's infrastructural expansion at Google Cloud's Next conference in San Francisco on March 8, 2017. See "Google Cloud Next '17 – Day 1 Keynote," March 8, 2017, 2 h., 4 min., 35 sec., posted by Google Cloud, Youtube, https://youtu.be/j_K1YoMHpbk.

the spatial complexities of their construction. This remapping thus necessitates a geographic recategorization of places that produce, support, and maintain the smart city and its operations.

Instruments of Territorial Expansion

If the factory was the architectural embodiment of industrialization, and the downtown tower stood in for the finance economy, the data centre is the quintessential building type of the contemporary data economy. However, the significance of data centres goes well beyond the formal and the symbolic. Data centres can also be conceptualized as instruments of territorial expansion critical to the extension of the practices and processes of smart urbanism. Our increasing dependence on mobile devices and the expansion of sensors and extractive data devices in everyday spaces have translated into massive amounts of data generated daily. In 2018, Cisco's Global Cloud Index forecasted that the amount of annual global data centre traffic will grow three-fold from 2016 to 2021, reaching 19.5 zettabytes per year by the end of 2021. (Cisco 2018)[2] All this data needs to be captured, stored, and processed for it to have any value for platforms. This is the function of data centres. In parallel, while mobile devices like smartphones and tablets are increasingly powerful, they rely heavily on the power of computing resources centralized in data centres to perform the complex tasks we demand of them every day. Belying the seemingly incredible power they put in the palms of our hands, the processing capacities of these devices are mainly directed towards input and output functions. Broadband networks ensure the near-instantaneous communication of the results from the data centre to the device. In this regard, personal devices are more akin to the terminals of timesharing mainframes than completely localized personal computers.

Reliability of access to this data cache is an important factor within the operations of the smart city. Building up redundancies in the network through the creation of multiple data regions and zones is an essential part of making a data platform reliable. This is also a spatial project, as regions and zones of the cloud are materialized through data centres, which are in turn grounded in the specificities of geography and location. There is a reciprocal relationship between the growth of data centres as storage and processing epicentres of the cloud and the massive rise of personal mobile computing and the extension of computing capacities

---

2 As a reference: 1 zettabyte = 1,000 exabytes = 1 million petabytes = 1 billion terabytes = 1 trillion gigabytes.

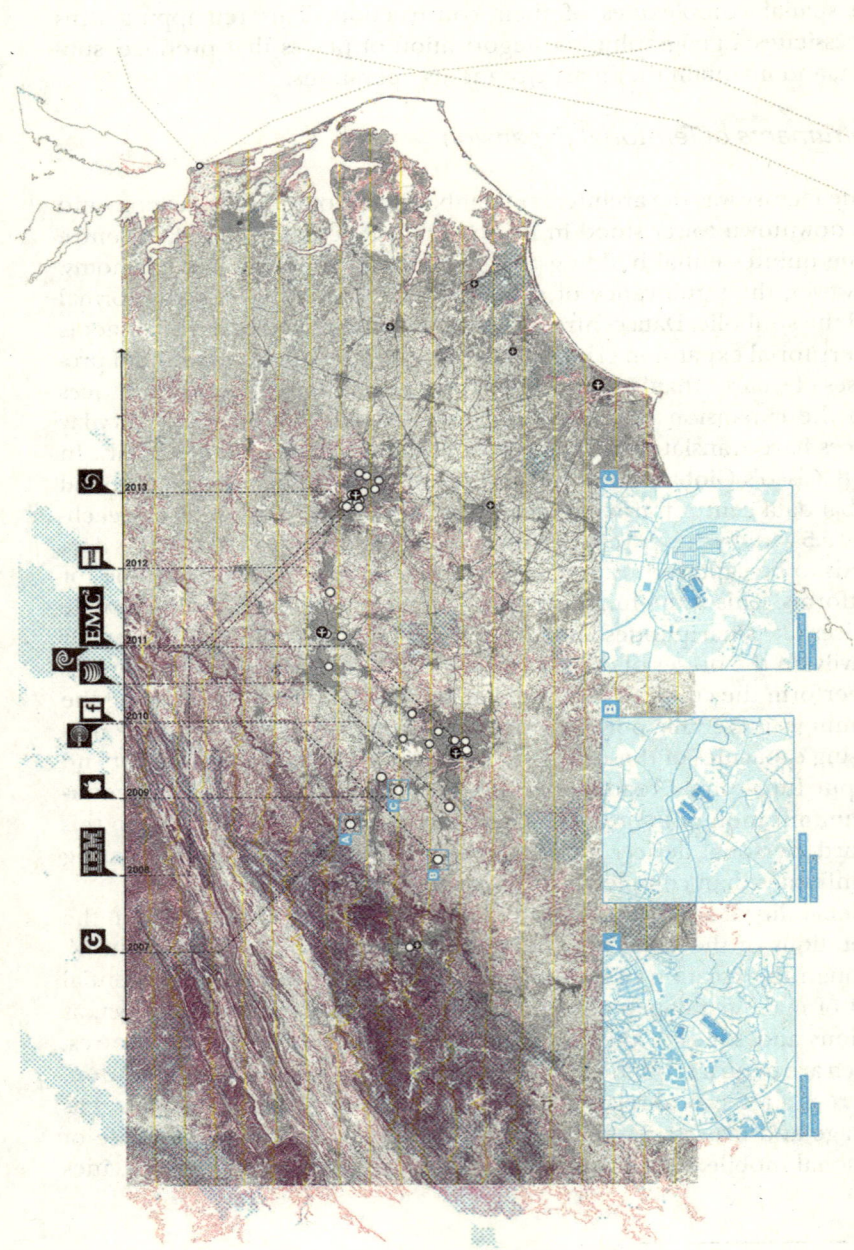

Figure 13.1. Google's Lenoir data centre in North Carolina – one of the first purpose-built data centres constructed by the company – sparked the growth of a data centre geography in the state.

Reframing Smart Urbanism 143

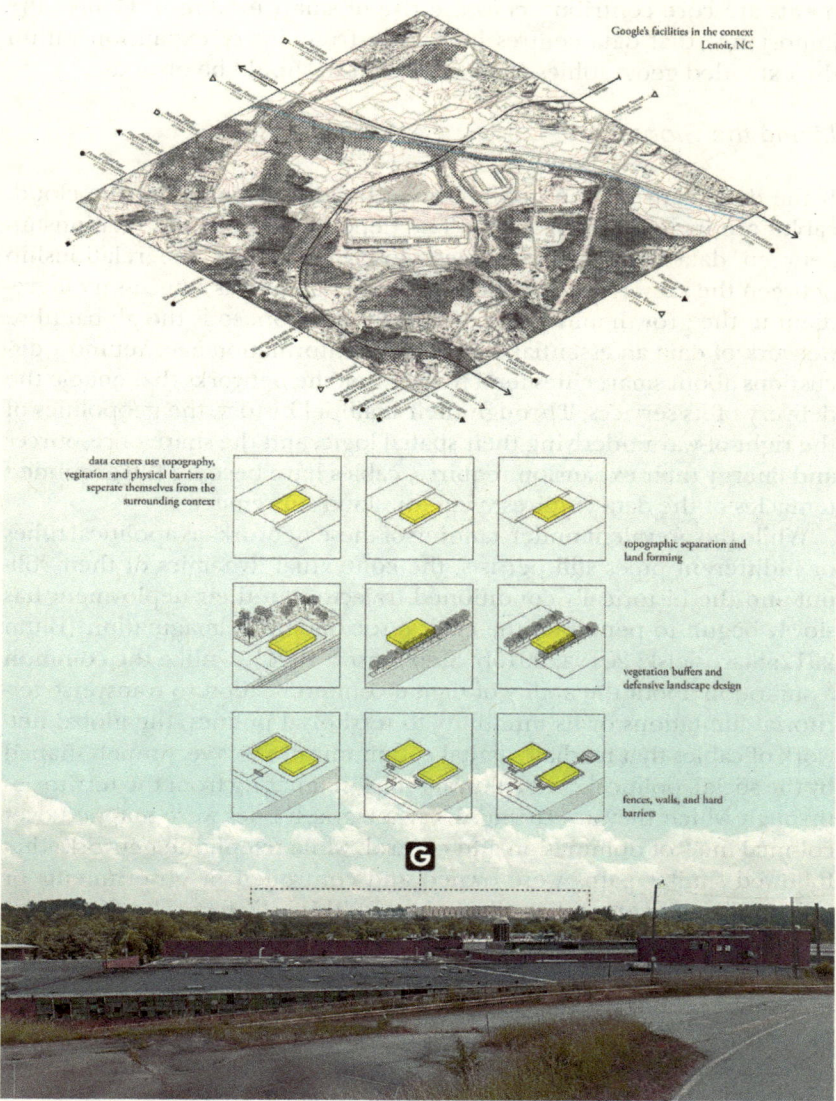

Figure 13.2. Google's Lenoir facilities in context. Bottom: Google's data centres perched on top of the town amid abandoned factories. Middle: typical spatial strategies of data centres. Top: Lenoir data centres in the context of their proximities and spatial dependencies.

to everyday spaces and infrastructures of urbanization. Both developments are core contributors to the rise of smart urbanism. Hence, the importance that data centres hold as instruments of expansion within the extended geographies of smart urbanism should be obvious.

Lining the Globe

If the data centre is the quintessential building typology of the cloud, cables are its logistical network. The dependency and the relationship between data centres and cables closely resembles the relationship between the railway and the factory. As the railway was a necessary ingredient in the growth and reach of industrialization, so is the global fibre network of data an essential aspect of the information age. Yet most discussions about smart cities tend to overlook the networks that enable the delivery of its services. Through their material history, the geopolitics of the right-of-way underlying their spatial logic, and the share of resources and energy their expansion requires, cables have become the grounded tentacles of the data empires vying for cloud supremacy.

While the prevalent understanding of these networks as apolitical tubes or indifferent pipes still persists, the contextual dynamics of their rollout and the historically conditioned trajectory of their deployment has slowly begun to penetrate the critical sociopolitical imagination (Blum 2012; Starosielski 2015a, 2015b; Stephenson 1996). Unlike the common assumption about the ability of digital communication to transverse territorial limitations or its immunity to territorial politics, the global network of cables that mediate digital communications is very much shaped by the social, political, and cultural forces emanating from the territories through which they are strung. Early telegraph lines were mapped over colonial lines of domination and control, while telephone networks that followed similar paths were owned and controlled by governments or state-affiliated monopolies (Starosielski 2015b, 27–63). The fibre-optic lines of the internet's backbone were similarly threaded along the very same "historical and political lines, tending to reinforce existing global inequalities," and for the most part have contributed to the continuity of historical power relations (Starosielski 2015b, 12). The contemporary network is largely colocated with other infrastructures of railways, roads, and pipelines (Durairajan et al. 2015).

The continuity of operations that accompanies the infrastructural coupling of communications on previous networks is not solely limited to the physicality of their lines. These networks also tend to inherit the sociopolitical messiness and the techno-ideological determinism of the host networks. As Ingrid Burrington writes about fibre-optic networks, "[they] tend to follow networks, and telecommunications and transportation networks tend to end up piled on top of each other" (2015). While the historical trajectory of the fibre networks is not always easily recognizable, "it's there, forming a kind of infrastructural palimpsest, with new technologies to annihilate space and time inheriting the idealized promise and the political messiness of their predecessors" (Burrington 2015). It is on this "infrastructural palimpsest" that the most recent transformation of the global networks of data is unfolding.

Extraction in the Cloud

While mobile devices, sensors, and urban monitoring instruments play an important part in the extended geography of smart urbanism, they are also material assemblies. And as such they connect the operations of smart cities to the more traditional sites of resource extraction (i.e., mines). Consider an Apple device, such as its flagship iPhone. The phone is designed in California by large teams of engineers, product designers, interface designers, and user experience specialists. However, the supply chain that extends beyond the Apple campus is expansive. The phones are assembled in Foxconn plants in Zhengzhou, Northern China, under questionable labour practices (Dou 2016). The parts that are assembled in Foxconn plants are sourced from the United States, Asia, and Europe. The assembled phones themselves are sent to intermediate warehouses at UPS or FedEx for online customers, or to Apple's Elk Grove, California, facilities for retail stores and other distributors (Supply Chain 24/7 2013).

The minerals and the elements that go into each component of a mobile device come from all over the world, sometimes from the most troubled regions. About half of the global supply of cobalt, which is a key component in lithium-ion rechargeable batteries used widely in smartphones and tablets, comes from the Democratic Republic of the Congo (DRC). A 2016 report by Amnesty International found that of the total cobalt exported from DRC, 20 per cent come from artisanal mines with significant human rights issues, including the use of child labour (Amnesty International 2016, 4). A further inquiry by *The Washington Post* confirmed the prevalence of cobalt from artisanal mines that use child labour in the global supply chain of cobalt (Frankel 2016). A single Apple iPhone uses 6.59 grams of cobalt, which accounts for more than 5 per cent of the total weight of the device. There are only six raw elements that are used more than cobalt in the iPhone: aluminum (24.14 per cent), carbon (15.39 per cent), oxygen (14.5 per cent), iron (14.44 per cent), silicon (6.31 per cent), and copper (6.08 per cent) (Merchant

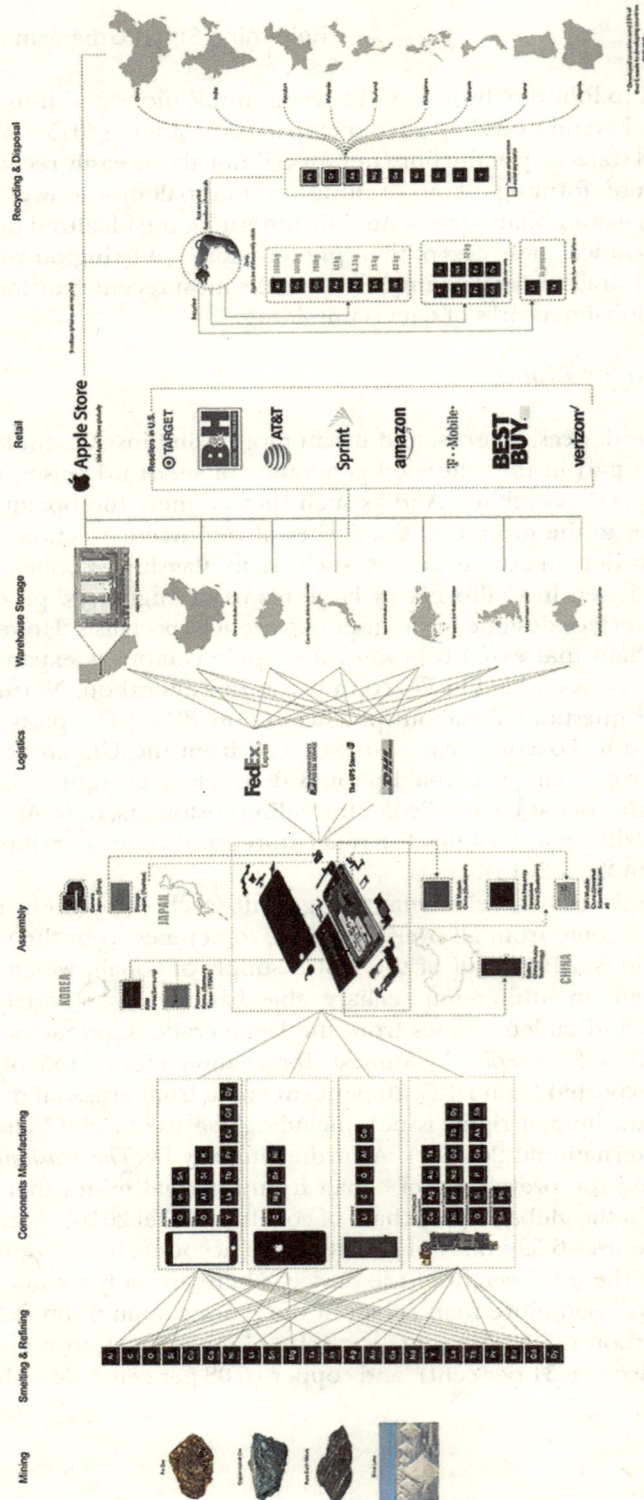

Figure 13.3. The life and death of mobile devices: from mining and refining of the various components, to assembly, logistics, recycling, and ultimately waste.

2017). With the global shipment of smartphones growing to 1.2 billion units in 2024, one can imagine the massive amount of cobalt used in these units (Sherif 2025). Once we consider the thirty or so other elements that make up the iPhone, the reach of the productive landscape of mobile computing touches almost every corner of the globe (Apple 2023). If we extend our understanding of the footprint of mobile devices beyond the glass walls of the Apple store and into the variegated landscapes of extraction, labour, energy, and waste that is produced within the production and supply chain of mobile devices, the scope becomes undeniably global.

Reframing Smart Urbanism

There is tremendous value in considering the operational spaces of smart urbanism along with its concentrated forms in cities. Understanding how data platforms at the core of smart urbanism operate is an important first step in reframing our understanding of smart urbanism and how the notion of smartness is materially constructed and supported. Ultimately, through this reframing, we can ground the exuberant promises of the smart city within the material realities of its production. We can position the smart city as a technologically driven form of neoliberal urbanism that operates through enclosure, privatization, and territorial expansion, instead of the revolutionary new urban paradigm that it is marketed as. And by complicating the spatial geography of tech platforms in this way, we can eventually peel back the transcendent veneer of tech urbanism to reveal how data clouds are firmly tethered to the messy geopolitics and the uneven material geographies of global urbanization.

Image Credits

All graphics and images by the author.

References

Amnesty International. 2016. *Democratic Republic of Congo: "This Is What We Die for": Human Rights Abuses in the Democratic Republic of the Congo Power the Global Trade in Cobalt.* Amnesty International. www.amnesty.org/en/documents/afr62/3183/2016/en/.
Apple. 2023. *Smelter and Refiner List.* Apple Inc. https://www.apple.com/nz/supplier-responsibility/pdf/Apple-Smelter-and-Refiner-List.pdf.
Blum, Andrew. 2012. *Tubes: A Journey to the Center of the Internet.* Ecco/HarperCollins.

Burrell, Jenna. 2020. "On Half-Built Assemblages: Waiting for a Data Center in Prineville, Oregon." *Engaging Science, Technology, and Society* 6 (June): 283–305. https://doi.org/10.17351/ests2020.447.

Burrington, Ingrid. 2015. "How Railroad History Shaped Internet History." *Atlantic*, November 24. www.theatlantic.com/technology/archive/2015/11/how-railroad-history-shaped-internet-history/417414/.

Cisco. 2018. "Global Cloud Index Projects Cloud Traffic to Represent 95 Percent of Total Data Center Traffic by 2021." https://newsroom.cisco.com/c/r/newsroom/en/us/a/y2018/m02/global-cloud-index-projects-cloud-traffic-to-represent-95-percent-of-total-data-center-traffic-by-2021.html.

Dou, Eva. 2016. "Deaths of Foxconn Employees Highlight Pressures Faced by China's Factory Workers." *Wall Street Journal*, August 21. www.wsj.com/articles/deaths-of-foxconn-employees-highlight-pressures-faced-by-chinas-factory-workers-1471796417.

Durairajan, Ramakrishnan, Paul Barford, Joel Sommers, and Walter Willinger. 2015. "InterTubes: A Study of the US Long-Haul Fiber-Optic Infrastructure." In *Proceedings of the 2015 ACM Conference on Special Interest Group on Data Communication – SIGCOMM '15*. ACM Press. https://doi.org/10.1145/2785956.2787499.

Frankel, Todd C. 2016. "The Cobalt Pipeline." *Washington Post*, September 30. www.washingtonpost.com/graphics/business/batteries/congo-cobalt-mining-for-lithium-ion-battery/.

Greenfield, Adam. 2017. *Radical Technologies: The Design of Everyday Life*. Verso.

Halpern, Orit, Jesse LeCavalier, Nerea Calvillo, and Wolfgang Pietsch. 2013. "Test-Bed Urbanism." *Public Culture* 25 (2): 272–306. https://doi.org/10.1215/08992363-2020602.

Harvey, David. 2001. "Globalization and the 'Spatial Fix.'" *Geographische Revue* 3 (2): 23–30.

Jasanoff, Sheila. 2004. *States of Knowledge: The Co-Production of Science and Social Order*. Routledge. https://doi.org/10.4324/9780203413845.

— 2015. "Future Imperfect: Science, Technology, and the Imaginations of Modernity." In *Dreamscapes of Modernity: Sociotechnical Imaginaries and the Fabrication of Power*, edited by Sheila Jasanoff and Sang-Hyun Kim. The University of Chicago Press. https://doi.org/10.7208/chicago/9780226276663.003.0001.

Koolhass, Rem. 2015. "The Smart Landscape." *Artforum International* 53 (8): 212, 215–16. https://www.artforum.com/features/the-smart-landscape-intelligent-architecture-223415/.

Mauldin, Alan. 2016. "Rising Tide: Content Providers' Investment in Submarine Cables Continues." *TeleGeography* (blog), May 27. https://blog.telegeography.com/rising-tide-content-providers-investment-in-submarine-cables-continues.

Merchant, Brian. 2017. "Everything That's Inside Your iPhone." *Vice*, August 15. www.vice.com/en_us/article/433wyq/everything-thats-inside-your-iphone.

Miller, Rich. 2021. "AWS Has Spent $35 Billion on Its Northern Virginia Data Centers." *Data Center Frontier*, October 4. https://datacenterfrontier.com/aws-has-spent-35-billion-on-its-northern-virginia-data-centers/.

Picon, Antoine, ed. 2015. *Smart Cities: A Spatialised Intelligence*. Wiley. https://doi.org/10.1002/9781119075615.

Sherif, Ahmed. 2025. "Global Smartphone Shipments from 2009 to 2024." Statista. www.statista.com/statistics/271491/worldwide-shipments-of-smartphones-since-2009/.

Srnicek, Nick. 2017. *Platform Capitalism*. Polity Press.

Starosielski, Nicole. 2015a. "Fixed Flow: Undersea Cables as Media Infrastructure." In *Signal Traffic: Critical Studies of Media Infrastructures*, edited by Lisa Parks and Nicole Starosielski. University of Illinois Press.

— 2015b. *The Undersea Network*. Duke University Press. https://doi.org/10.2307/j.ctv11smhj2.

Stephenson, Neal. 1996. "Mother Earth Mother Board." *Wired*, December 1, https://www.wired.com/1996/12/ffglass/.

Supply Chain 24/7. 2013. "Is Apple's Supply Chain Really the No. 1? A Case Study." Supply Chain 24/7, September 2. www.supplychain247.com/article/is_apples_supply_chain_really_the_no._1_a_case_study.

14 The Smart Oasis: Smartification as Process

SEBASTIAN BORNSCHLEGL

Artist Statement

Where does the smart city begin and end? How do smart infrastructures and sensing technologies demarcate its boundaries, and how is urban life reshaped within it? These questions characterize global smartification processes that transform urban spaces through incremental changes and sudden recreations. As Halpern and Mitchell (2023) argue, smartness should be understood as a specific kind of epistemology: a particular way of collecting knowledge about the world as data, transforming it through algorithmic means, and making this data actionable in an automated way for various purposes. The smartification of the urban fabric encompasses its reconfiguration through the integration of digital sensors, automatons, surveillance techniques, and new relationships between non-human actors and human subjects inhabiting the smart city.

I explore such a reassembly at the local level using the case of Seestadt, Vienna's flagship urban development project. On the site of a former airfield at the edge of Donaustadt, a "city within a city" has been built over the last decade.[1] It is accompanied by smart city storytelling (Söderström et al. 2014) that promises a modern urban lifestyle made possible by state-of-the-art urban design, smart infrastructure, and innovative business opportunities. As Douglas-Jones points out in the foreword, smartification imbues the mundane aspects of urban life with the shine of a better future based on the dream of perfect technological integration. This glare of the smart city can be blinding for critical reflection on who

---

1 Die Seestadt Wiens, "Urban Lab," aspern, 2024, www.aspern-seestadt.at/en/business_hub/innovation__quality/urban_lab.

these futures actually serve and what sociotechnical dependencies they are shackled to.

In my visual vignette (Gugganig and Douglas-Jones 2021), I approach Seestadt through the metaphor of the smart oasis surrounded by the urban analogue desert. It captures the discursive and material processes inherent in smartification: the drawing of boundaries around a seemingly privileged and superior patch of the urban fabric, set apart from the urban legacy that is implicitly or explicitly framed as deficient. Given the fragmented nature of urban infrastructures (Graham and Marvin 2001), I challenge this kind of prophecy by exploring the actual reach and boundaries of Seestadt's smart infrastructures. Methodologically, this was done by literally following different kinds of sensors belonging to two mobility services around the neighbourhood: the former test operation of two semiautomated buses and a docked bicycle-sharing system.

The results of this auto-ethnographic work reveal how smartification, as a global phenomenon, has been specifically adapted to Seestadt and what kinds of transformation its epistemology entails. The test operation of the automated vehicles (AVs) illustrates how the digital gaze of laser beams and cameras creates an affective atmosphere that connects "presences and absences, the visible and invisible" (Tironi and Valerrama 2021, 54). The AVs encapsulated road users along their short route in an experiment of what can be seen and understood digitally and automatically, potentially triggering boredom, excitement, or fear depending on the individual's subject position. Interacting with the shared bikes of SeestadtFLOTTE exemplifies the ontogenetic qualities of sensing technologies, requiring users to adapt new "modalities of perception and interaction" and "remodulat[ing] patterns of attention" (Antenucchi 2021, 93) to the required authentication device, the availability of docking stations and bikes, and the geographic limitation of the service.

Looking at the boundaries and integration of the smart infrastructures studied, they seem rather limited: automated buses that are hardly usable as a practical means of transportation; a bike-sharing system whose stations are limited to Seestadt, complementing rather than providing essential mobility services. The takeaway is that Seestadt is not a uniform smart city that encapsulates its inhabitants. Rather, it is divided into multiple small oases of smartness that are only partially connected and unable to form a coherent boundary. This aspect is reflected in the multilayered collages of the vignette. Nevertheless, the housing estate is indeed an oasis for its inhabitants. Its prosperity seems to be linked primarily to thoughtful urban planning, a focus on public transport, and high-quality urban design. These features are neither limited to Seestadt nor tied to cutting-edge technology, challenging us to rethink what constitutes smart infrastructure and urban development. For example, compared to the disruptive and profit-driven operations of dockless scooter-sharing services (Shepherd and Walker, this volume), bike-sharing in Seestadt may not be as novel or convenient. But overall, the SeestadtFLOTTE provides a safer, more affordable, and more accessible public service.

The Visual Vignette *The Smart Oasis: Smartification as Process* can be accessed at https://archive.org/embed/visual-vignette-seestadt.

References

Antenucchi, I. 2021. "Smart Cities, Smart Borders: Sensing Networks and Security in the Urban Space." In *Sensing in/security: Sensors as Transnational Security Infrastructures*, edited by Nina Klimburg-Witjes, Nikolaus Poechhacker, and Geoffrey C. Bowker. Mattering Press. http://doi.org/10.28938/9781912729111.

Graham, Stephen, and Simon Marvin. 2001. *Splintering Urbanism.* Routledge. https://doi.org/10.4324/9780203452202.

Gugganig, Mascha, and Rachel Douglas-Jones. 2021. "Visual Vignettes." In *Sensing in/security: Sensors as Transnational Security Infrastructures*, edited by Nina Klimburg-Witjes, Nikolaus Poechhacker, and Geoffrey C. Bowker. Mattering Press. http://doi.org/10.28938/9781912729111.

Halpern, Orit, and Robert Mitchell. 2023. *The Smartness Mandate.* MIT Press. https://doi.org/10.7551/mitpress/14623.001.0001.

Söderström, Ola, Till Paasche, and Francisco Klauser. 2014. "Smart Cities as Corporate Storytelling." *City* 18 (3): 307–20. https://doi.org/10.1080/13604813.2014.906716.

Tironi, N., and M. Valerrama. 2021. "Microclimates of (In)Security in Santiago: Sensors, Sensing and Sensations." In *Sensing in/security: Sensors as Transnational Security Infrastructures*, edited by Nina Klimburg-Witjes, Nikolaus Poechhacker, and Geoffrey C. Bowker. Mattering Press. http://doi.org/10.28938/9781912729111.

15 Disposable: Infrastructures of Exclusion in Dharamshala's Smart City

HANNAH CARLAN

In 2015, the current prime minister of India, Narendra Modi, launched the "Smart Cities Mission": an urban development project aimed at bringing sustainable infrastructure to one hundred cities across India. The Smart Cities Mission involves a wide array of development initiatives, ranging from digital governance portals and electric buses to public parks and recreation facilities. In this essay, I examine how one such city, Dharamshala, became "smart" in the foothills of the western Himalayas. Drawing upon fieldwork conducted in 2017–22, I examine how the boundaries of smartness were interpreted through long-standing hierarchies of class, caste, and gender, reinforcing inequalities that threatened the lives of those deemed external to it. Through ethnographic analysis of the process of defining the borders of the smart city, implementing underground dustbins, and managing waste dumping sites, I reveal how the promise of "smartness" in Dharamshala generated a *necropolitics of disposability* through which certain human and non-human lives were rendered expendable. This exploration reveals how smartification generates social infrastructures of exclusion.

The Semiotics of Smartness

Smart development paradigms are often framed as mitigating the effects of unsustainable urban growth globally by mobilizing climate-friendly and inclusive technologies. However, the meanings of "smartness" cannot be presumed across contexts and should be investigated ethnographically to understand how transnationally salient discourses articulate with long-standing ideologies, histories, and inequalities. Anthropologists have long sought to understand how development paradigms interact with existing social and environmental contexts, producing frictions that often generate unforeseen consequences (Li 2007; Tsing 2005). Central to this disciplinary tradition is the recognition that transnational discourses – like "smartness" – necessarily transform, and are transformed by, the everyday practices and power dynamics of diverse social worlds. Ethnography thus allows us to investigate how the meanings of "smartness" are constituted by everyday actors, and what long-standing inequities may be reinscribed through seemingly novel discourses.

In Dharamshala, smartness became synonymous with "cleanliness," which carried additional distinctions of caste and class privilege. In doing so, those human and non-human entities seen as "dirty" were then rendered external to smartness, and thus disposable. Constituting the meanings of smartness was thus centrally about carving out the boundaries of smartness, and in doing so, enacting its limits. Recently, anthropologists have attended to the role of urban development initiatives in generating new material and social infrastructures of waste, which are embedded in deeply unequal hierarchies of class, caste, race, and gender (Ahmann 2019; Butt 2020). Smart development not only requires disposing of large amounts of physical waste, but it also makes disposable certain life forms that are rendered outside the purview of the state and the market – informal workers, settlements, and economies (Sanyal 2007). Such exclusions are not new to smart-city building but are continuations of governmentality through exclusion, in which those outside the purview of citizenship are deemed unworthy of care (Ong 2006). Investigating the bounded semiotics of smartness across contexts thus requires attending to the exclusions generated therefrom and their embeddedness in long-standing histories of inequity.

In the remainder of this chapter, I explore how making Dharamshala "smart" required implementing a locally constituted and contested regime of policy and infrastructure changes that enacted the boundaries of smartness by making specific places, people, and animals disposable. I offer ethnographic examples gathered from interviews and observations collected from 2017 to 2019, and again in 2022, as part of a larger project that examined the politics of rural development in Dharamshala's state of Himachal Pradesh (Carlan 2021).

Making the City "Liveable"

In 2015, Dharamshala won a bid to be one of one hundred cities in India to receive a five-year grant from the central and state governments to overhaul its urban infrastructure and governance systems. To qualify for the funding, the small town of around twenty thousand residents was expanded to include five surrounding villages, previously governed by rural, independent village councils, nearly doubling the population overnight. The city was upgraded from a municipal council (*nagar parishad*) to a municipal corporation (*nagar nigam*), expanding from ten to seventeen wards. Residents are represented by ward councillors, along with a mayor and deputy mayor, who retain nominal powers of decision-making and whose roles are largely ceremonial. The municipal commissioner, who is appointed by the central government, oversees all executive and administrative duties, and in the case of smart cities, is also its CEO. Smart cities in India are administered by "special purpose vehicles": privatized limited companies that solely oversee the planning and execution of all development projects within the city, which are controlled by a CEO and appointed board of directors. As such, smart city development projects are entirely distinct from public participation or elections within the municipal system, running in a corporate model in which public participation is largely immaterial.

In October 2018, I interviewed the then-commissioner of the municipal corporation of Dharamshala and the CEO of the smart city, Sandeep Kadam. We spoke for three hours about the challenges facing the smart city project. When I asked him what makes a city "smart," he said: "We are introducing something new, which makes the city more liveable. The outcome of that effort is a smart city. For example, with a smart street, you can add something to the street that makes it more usable for the citizens." In his narrative, the municipal commissioner painted a broad picture of what smartness means for the development of roads, which he emphasizes from the beginning as being focused on improving the "liveability" and "usability" of the city for local citizens. He went on to explain that the amenities and infrastructure that would make the city more usable for citizens – including large footpaths, cycle lanes, street furniture, lighting, water fountains, etc. – are largely impossible due to the city's small, narrow roads and hilly geography. Kadam continued:

> By concept, smart roads promote walking [rather] than driving. Now, in Dharamshala, we are not making this exactly because to make this you need very wide roads. We don't have that. We have very small roads. So basically, what we are doing is [making] a very good surface and underground amenities. This is our concept of "smart road."

What urban "liveability" entails is not only context-dependent and highly contested, but shifts within everyday narratives as people grapple with the implications of rapid urban change (Lam-Knott 2022). Throughout his narrative, the municipal commissioner offers an attenuated meaning of smartness, one that selectively retains some aspects over others. In the example he offers of a smart road, it is underground amenities and a good surface – those which are most conducive to driving – which are prioritized over above-ground accessibility. As such, both "liveability" and "usability" are framed here as explicitly aligned with the needs of drivers rather than pedestrians. This interpretive move is not framed as a political choice but rather one that is determined by the pre-existing geographical conditions of the city's narrow, winding roads. Still, the commissioner's interpretation of smartness implicitly centres the needs of drivers and tourists who flood the region, which he frames as a definitively localized idiom of smartness – this is "*our* concept of smart road." The commissioner's statements demonstrate that smartness is not a decontextualized, abstract model but rather is a continuously produced and contested project of reconciling ideal visions with actual conditions and realities.

Rather than merely attribute the commissioner's localized meaning of smartness to the unequal consequences of development, I instead argue that such interpretations of smartness are themselves constitutive of development. In Dharamshala, making the city more "liveable" required making certain aspects of the city *disposable*, including specific people, places, and animals. Disposability is not merely an unintended negative consequence of poorly planned development, but rather it is inherent to the production and coherence of smartness, and of development more broadly. Making the city "smart" required implementing certain projects while eliminating those seen as unfeasible; the former often introduced new infrastructures that were in fact not usable for a majority of the population, and in many cases made the city *less* liveable for them. The question is thus not what becomes expendable through development but rather how such expendability becomes constitutive of development.

Hodgepodge: Mapping the Smart City

The first step in making Dharamshala smart required redrawing the city's borders, transforming the surrounding rural village councils (or *panchayats*) into city wards in order to meet the population threshold of 50,000 residents required to qualify for the Smart Cities Mission. Since the municipal corporation was going to impose a sweeping array of new taxes for its urban amenities, including water, electricity, and sanitation, along with new property taxes, those who sought to avoid such financial burdens were able to lobby for the exclusion of their villages from the newly constituted smart city. There was widespread resistance to the smart city project among local residents who feared the costs of living in a highly taxed urban environment that would eliminate the many welfare benefits provided to rural citizens in India; however, such people were largely powerless to oppose it outright. For those elites who had long-standing political power, however, there were avenues through which to influence the shape that the city would take.

Mamta, the headperson (*pradhan*) of a rural village council neighbouring Dharamshala, explained in an interview in 2018 that when the municipal corporation was established, there was a great deal of confusion about which village areas would be included and which would be excluded. "I don't know what method they used to decide which areas to cut, but what they ended up doing was keeping in the areas with poor people and leaving the rich folks out." Her own village council area (*panchayat*) was suddenly divided in half – with the poorer areas being incorporated into the smart city and the wealthier areas remaining in the rural *panchayat*, thus untaxable by the municipal corporation. I heard a similar account from Vishal, ward councillor for a neighbouring village that had been incorporated in 2015. He explained that the boundaries of the corporation were drawn by the members of the ruling party in Dharamshala, the Congress Party, who had organized the survey of households. "They did the survey themselves, [saying] 'include this house, leave that one out.'" The result, he said, was "pure confusion" about where the city began and ended:

> You won't believe that even six months after I was elected [as ward councillor], I still didn't know where my area began and where it ended, or how many households there were. People told me afterwards that when they were organizing the [survey] meetings, people didn't attend. They didn't know. It was a complete hodgepodge, pure confusion. And the really well-off families, the ones who were financially strong, they were left out of the municipal corporation. And the smaller, poorer families, they were brought in. At that time, I was completely shocked and had no idea who was doing the [decision-making] and who was having it done.

Both Mamta and Vishal, despite their different positionalities as a rural, female village headperson and an urban, male ward councillor, similarly reported that the process of drawing the boundaries of the corporation was extremely politicized. Wealthy residents were able to exclude their neighbourhoods from incorporation in order to avoid the increased burdens of taxation for their properties and amenities. This included the home of Dharamshala's then-legislative assembly representative, Sudhir Sharma, who was responsible for securing the bid for the smart city project during his tenure from 2003 to 2017. Sharma's newly constructed, impressive home is located in a village that was excluded from the boundaries of the municipal corporation, despite being bordered by several other villages that were included. The map itself thus became a visual icon of the "hodgepodge" and "pure confusion" through which the smart city was founded, in which the need to increase the population came at the expense of poor citizens' input, who were suddenly subject to an unfamiliar and burdensome system of governance and taxation.

"Cleansing" the City

Upon the very formation of the smart city, bureaucratic processes of eviction laid bare the exclusionary nature of smartness in Dharamshala. In early 2016, a community of 1,500 migrant workers who had been living in the heart of Dharamshala for three decades were evicted and their housing structures were razed under the auspices of improving "cleanliness" and ending "open defecation" in the smart city (Asher 2017). The residents were primarily members of Indigenous communities from other states in northern India, who had migrated to work in construction and other manual labour jobs and had been building much of the city's infrastructure for decades.

The municipal commissioner, Sandeep Kadam, offered a bureaucratic justification for the eviction: he said that the area was destroyed because it was a "non-notified slum." When I asked him whose responsibility it is to "notify" a slum, he said that he was avoiding this point and would rather not discuss it further. Others were more explicit in their justification: many said that its residents were "dirty" and "diseased" and were polluting local water sources with open defecation. However, the timing of the eviction – at the very beginning of the smart city project – belied the political motivation for the dramatic move. The migrant workers' settlement was located on what activist Manshi Asher described in an interview in 2018 as "prime land" directly in front of Dharamshala's famed cricket stadium. Asher was clear that the move's official justification – to end open defecation – was in fact a strategic move to make way for planned development in that area. Shortly after the community was displaced, the state of Himachal Pradesh was declared India's second "Open Defecation Free" state and received a massive loan from the World Bank. The displacement of this community was widely criticized by activists, who pointed out the discriminatory exclusion of migrants from the smart city (Asher 2017; Nigar and Mahar 2016). Making the city smart did not lead to the provision of sanitation facilities for existing long-term residents but "cleansing" those residents entirely from its physical borders, destroying their housing and leaving them further vulnerable. As such, the transformation of Dharamshala into a smart city required disposing of undesirable citizens, and thus, "instead of addressing existing

Figure 15.1. Sensor-based underground dustbins located throughout the Dharamshala Municipal Corporation. Photo by author.

social exclusions, [it] actually reinforce[d] long-standing social inequalities" (Datta 2015, 4).

Smart Graveyards

With the newly drawn borders of the city, there was a two-fold increase in the population nearly overnight in 2015. One of the first challenges faced was in expanding the municipal sanitation service. Waste management not only generated a quite literal politics of disposability for local residents but also for non-human beings living within and beyond the boundaries of the city. In 2017, the city installed over two hundred underground "smart" sensor-based dustbins featuring separate organic and inorganic underground repositories that alert authorities when they are full. After these dustbins were installed, the municipal corporation outsourced waste management to private contractors, who did not regularly empty them, and all of the dustbins quickly fell into disuse and various states of overflow and disrepair (Mohan 2019).

As a result of the dustbins' early failure, these sites became hotbeds of street pollution, piling up and attracting a vast array of wild animals, including dogs, cows, and monkeys. Many of these animals would often wander into the neighbouring homes and fields in search of food to cause additional destruction to homes and crops. Dogs were the primary casualties of the dustbins as they would often fall into the underground repositories and become unable to escape them. The dustbins became widely known as "animal graveyards." On one occasion, I chanced upon a smart dustbin with a group of friends who had heard the soft cries of a dog coming from it, and we opened it to find a dog curled up, ten feet below, atop the trash heap. I called the municipal commissioner on his cell phone and asked that he send someone to empty the garbage and release the dog, but he informed me that this was very common, and they usually pull out a dead dog when they empty the cans.

The fallout of the smart dustbins laid bare the necropolitics of smartification in Dharamshala, wherein making the city "liveable" for some generated the death and destruction of life for others. Residents of a neighbouring village, Sudherd, were further implicated in this necropolitics of disposability, as the dead animals extracted from dustbins were primarily dumped at the municipal waste management site, which was located on an illegal dumping site that fell within the boundaries of their village area (*panchayat*). The dumping site was itself a long-standing problem that residents of Sudherd had been contesting; however, with the now-doubled population, the municipal corporation had ramped up dumping at this site. This began to significantly pollute the land and water in Sudherd, causing residents to develop many new diseases, clogging up local drinking water sources, and degrading the fertility of the neighbouring farmland. Sunita, leader of a local women's collective (*mahila mandal*), pointed at the dumping site one day while we were walking through her village. "Dead dogs are thrown here, look over there, ten, fifteen, twenty dogs are there, and now the poor stray animals, they also eat the garbage there, so whatever dies, they bring it here and drop it. All of our land, our farm land, has become contaminated."

I worked closely with Sunita and her collective of women as they organized to protest the dumping site in 2018 and 2019. They had been lobbying the central government and the National Green Tribunal in Delhi, which adjudicates cases of environmental protection, to close the site and remove the mountain of garbage that had spread into their village for several years at that point. The village hamlet most proximate to the dumping site in Sudherd was home to a small group of Indigenous, or Scheduled Tribe, Gaddi families, a traditionally nomadic shepherding community that had become agricultural in recent decades. The women from these families showed me how the dumping site had clogged their local spring, which was their source of drinking water and irrigation,

Figure 15.2. A stray dog lies trapped at the bottom of a "smart" dustbin. Photo by author.

and how the ground underneath was no longer supporting decent crops for cultivation. Residents also told me stories of new forms of cancer, skin diseases, and other illnesses that villagers were experiencing. Sunita shared, "It's a very bad situation. Every month I don't know how many people are dying, from kidney failure, cancer, loose motions."

In addition to these health and livelihood challenges, many residents reported extreme social stigma and marginalization from their community networks as a result of the dumping site. People now refused to visit their homes or attend functions like marriages. One woman whose house is located at the direct base of the dumping site explained:

> We can't hold functions anymore. If we want to hold a function here and invite people, then how will we feed them? Flies, mosquitos, everything sits right on top of the food. So we can't invite anyone. People won't come. They say, "How can we go to their house? We've never eaten like that." But we *have* to eat. We are used to it. Where are we going to go? We have to eat here. Even if we die. If we are going to die, then we are going to die right here.

Sunita, herself an upper-caste, non-Indigenous woman, warned me before we entered the village area that I should not eat or drink anything while we were there (a directive I ignored), citing concerns about my health from consuming contaminated water. However, refusing water and food also indexes fear of caste-based pollution, which Sunita herself was inadvertently reproducing through her refusal to accept water from these women. Here, it is imperative to place the smart city's consequences within these longer-term histories of exclusion, as the dumping site was merely compounding an already extant caste-inflected rhetoric of pollution. Perhaps most disturbingly, the dumping site collapsed Sudherd's material conditions with its social ones, further entrenching modes of caste oppression (Butt 2020).

Conclusion: Infrastructures of Exclusion

When I returned to Dharamshala in July 2022, I was greeted by an unexpected addition to the landscape: a ten-foot metal fence had been installed along the road bordering the dumping site, painted bright green and emblazoned with Dharamshala's Smart City insignia. As I drove further toward Sudherd, I noticed another new feature of the landscape: the Dharamshala SkyWay had been completed, and an aerial tram now connected Dharamshala to the tourist centre and home of the Dalai Lama, McLeod Ganj. I watched from below as colourful compartments moved slowly along the suspended cables, moving tourists along a scenic path overlooking the Dhauladhar range. Situated directly below this scenery was the municipal dumping site, which had grown by at least a third of its size over the last three years. Later that day, I met with Sunita, who explained that COVID-19 had disrupted most of their organizing efforts, and that while their roadblock sit-in (*chakka jam*) in the main city centre in early 2020 was met with promises from municipal and district authorities to remove the dumping site, everything changed when India went into lockdown in March 2020. Since that time, previous Municipal Commissioner Sandeep Kadam had been reassigned, and the only visible change to the area was the fence that now rendered the dumping site invisible to residents of the smart city.

While many accounts of city- and state-making have highlighted the physical infrastructures – gates, walls, fences – that produce exclusion (Davis 2006; Sur 2021), Dharamshala's smartness was constituted first through a semiotic process of boundary-making. The city's shift to "smartness" hinged upon the production of certain places, people, and animals as disposable. Migrant labourers, Gaddi families, countless animals, and entire villages were thus extracted *within* the physical spaces

Figure 15.3. Dharamshala Municipal Corporation's dumping site sits atop the hill, above the farmland of several Gaddi families. Photo by author.

of smartness, recast as achievements of the city's project of sustainable development. The consequences of these moves – displacement, disease, and death – were temporally marked as pre-existing phenomena rather than the outcome of the smart city's creation and spread. Migrant encampments, the dumping site, and stray animals were problems that the municipal corporation had *inherited* from the earlier government, and thus their fallout was no fault of the smart city itself.

Whereas much scholarly critique has emphasized the uneven consequences of "smart" urban development, it is necessary to recognize that all forms of developmental governance are predicated upon selective exclusions, generated as actors constitute boundaries, extract resources, displace peoples, and reframe deservingness. In the case of the Smart Cities Mission in India, the goals of increasing "liveability" and "usability" make the inherent contradictions of such projects even more apparent as they continue to enact sociopolitical, environmental, and embodied infrastructures of exclusion, without delivering on their promise of sustainable infrastructure. The ironic interplay of "liveability" and death within Dharamshala's smartification project demonstrates that disposability is not a by-product of "smartness": it is its raison d'être. Given these contradictions, the analytical task for studies of smartification is to attend equally to the material infrastructures of city-making as well as to the semiotic infrastructures of meaning-making through which smartness is interpreted. Only then can we trace how smartness shifts across contexts, re-embedding long-standing forms of class, caste, and gender inequity.

References

Ahmann, Chloe. 2019. "Waste to Energy: Garbage Prospects and Subjunctive Politics in Late-Industrial Baltimore." *American Ethnologist* 46 (3): 328–42. https://doi.org/10.1111/amet.12792.

Asher, Manshi. 2017. "Is There No Place for the Poor in Dharamshala 'Smart' City?" Himdhara, September 17. www.himdhara.org/2017/09/17/is-there-no-place-for-the-poor-in-dharamshala-smart-city/.

Butt, Waqas H. 2020. "Waste Intimacies: Caste and the Unevenness of Life in Urban Pakistan." *American Ethnologist* 47 (3): 234–48. https://doi.org/10.1111/amet.12960.

Carlan, Hannah. 2021. "Producing Prosperity: Language and the Labor of Development in India's Western Himalayas." PhD diss., University of California, Los Angeles.

Datta, Ayona. 2015. "New Urban Utopias of Postcolonial India: 'Entrepreneurial Urbanization' in Dholera Smart City, Gujarat." *Dialogues in Human Geography* 5 (1): 3–22. https://doi.org/10.1177/2043820614565748.

Davis, Mike. 2006. *City of Quartz: Excavating the Future in Los Angeles.* Verso Books.

Lam-Knott, Sonia. 2022. "Contested Meanings of Urban Heritage in Hong Kong." *City & Society* 34 (1): 62–87. https://doi.org/10.1111/ciso.12420.

Li, Tania Murray. 2007. *The Will to Improve: Governmentality, Development, and the Practice of Politics.* Duke University Press Books. https://doi.org/10.1215/9780822389781.

Mohan, Lalit. 2019. "Sensor-Based Dustbins a Failure in Dharamsala." *Tribune*, December 25. www.tribuneindia.com/news/himachal/sensor-based-dustbins-a-failure-in-dharamsala-18810.

Nigar, Shazia, and Sumit Mahar. 2016. "Photo Story: Inhabitants of Charan Khad Slums Bear Brunt of Dharamshala Smart City Dream." *Wire*, October 21. https://thewire.in/rights/dharamashala-slum-smart-city.

Ong, Aihwa. 2006. *Neoliberalism as Exception: Mutations in Citizenship and Sovereignty.* Duke University Press. https://doi.org/10.1515/9780822387879.

Sanyal, Kalyan. 2007. *Rethinking Capitalist Development: Primitive Accumulation, Governmentality and Post-Colonial Capitalism.* Routledge India. https://doi.org/10.4324/9781315767321.

Sur, Malini. 2021. *Jungle Passports: Fences, Mobility, and Citizenship at the Northeast India-Bangladesh Border.* University of Pennsylvania Press. https://doi.org/10.9783/9780812297768.

Tsing, Anna Lowenhaupt. 2005. *Friction: An Ethnography of Global Connection.* Princeton University Press. https://doi.org/10.1515/9780691263526.

16 Wal*Smartification: Considering the Superimposition of Dockless Shared Electric Scooters on Fayetteville, Arkansas

DEVIN SHEPHERD AND JULIETTE WALKER

Artists' Statement

Shepherd and Walker's publication *Wal*Smartification: Considering the Superimposition of Dockless Shared Electric Scooters on Fayetteville, Arkansas* grew out of Shepherd's practice of photographing so-called dockless, shared electric scooters in Fayetteville, Arkansas, in 2021. Taking inspiration from the Smartification of Everything conference themes and through collaborating with Walker, whose recent projects have been publication/print-based and participatory, the scooter images – which continued to be collected by both collaborators into 2022 – were paired with an original essay and Risograph-printed as a booklet. The intention of this work was to document, examine, and describe in an accessible way how the privatization of public space occurs in seemingly novel and innocuous ways. In this case, it appeared as e-scooter dispersal by undemocratic means in the city where Shepherd and Walker lived.

Shepherd and Walker had previously combined images and text in a publication that focused on objects appearing in the environment and becoming unremarkable over extended periods of time. Their 2021 publication *Something to Notice* considers lobster traps littering the beaches of Maine in the midst of the COVID-19 pandemic, with Shepherd contributing words and images and Walker contributing images as well as printing and binding the final works. *Wal*Smartification* was produced in a similar manner and was completed with duties similarly distributed. In the essay portion, Shepherd describes how dockless, shared e-scooters are simply a recent example in a long-running tendency of businesses to cloak their privatization of public space in terms that valorize capital growth vis-à-vis appealing to beliefs about technological progress. In order to communicate the minutiae as well as the scale of this issue, the Northwest Arkansas region's most famous corporation, Walmart, is evoked. The invocation of smartness to achieve business aims is not only not new, Northwest Arkansas is arguably ground zero for the phenomenon, which the dockless, shared e-scooter companies are merely perpetuating. By highlighting the by-now-prosaic spectre of Walmart, a throughline to the present-day version of enclosure and appropriation under the guise of economic and societal progress via technology is drawn.

As Ali Fard discusses in this volume, the agents behind urban smartification, who tend to market the process of advancing smart urbanism in technocratic, techno-solutionist language, also elide and obscure the substrate of material effects such a process requires. For *Wal*Smartification*, the Risograph printing method was chosen both to mirror and contrast with the phenomenon of private scooter dispersal in public spaces. Risograph printing is minimally impactful on the environment as it uses plant-based inks and stencils – as opposed to toxic toner printers – arguably in extreme contrast to the superficially green-seeming scooters. Furthermore, printing this publication with a Risograph printer meant that the textual and photographic information it contains could be reproduced quickly, inexpensively, and at a large scale, potentially rivalling the "smart" e-scooters in the sense that the booklets too can be dispersed widely and with ease. Risograph ink, while permanent, is susceptible to touch and can be easily smeared even once set on the surface of the paper it is printed on. Furthermore, by creating a physical publication juxtaposing critical text with images of physical, found objects in public spaces – which might be considered, to use Hito Steyerl's term, "poor" or low-quality due to the heavy artefacting inherent in the process of Risograph printing digital photographs – *Wal*Smartification* draws attention to the inherent post-digitality of both the phenomenon the publication references and to the publication itself. In *Publishing as Artistic Practice* – a work that is foundational to Shepherd and Walker's approach to their work – Hannes Bajohr writes,

> If the digital is a concept of reality or a temporality, increasingly transparent to scrutiny, the post-digital is what performs the sudden yank that makes it apparent again. It provokes a disharmony in the structure of the obvious, thus drawing attention to it, and makes the process of technization experienceable. (2016, 104)

It is in this way that these digitally drafted, physically printed publications are intended to function. By mimicking the material and abstract qualities of the scooters, which frequently are experienced as physical impediments and impositions despite being supposedly ennobled and elevated by digitality, the booklets are intended to make palpable the discord between suppositions about smartness and actual lived experience.

While the prevailing narrative of smartification must omit and occlude details about the greater material impact of its implementation to be

further propagated, *Wal*Smartification* accentuates its materiality – through elements such as artefacted duotone images, typewriter typeface, and staple binding – in addition to problematizing the narrative that ostensible smartness is societally beneficial. The digital aspects of this work recede to emphasize its material qualities and the nature of its production in contrast with the subject matter being represented therein. Smartification, in part, Shepherd and Walker argue, depends

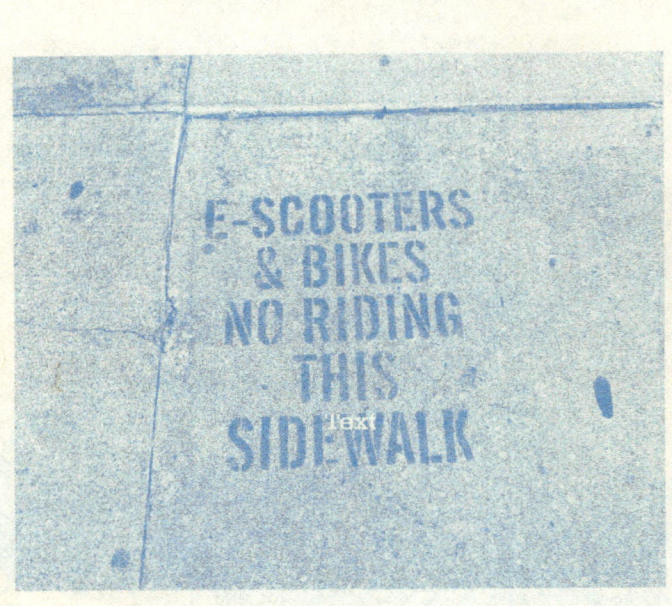

A collaborative publication made by
Devin Shepherd and Juliette Walker.

Writing – Devin Shepherd
Photographs – Devin Shepherd & Juliette Walker
Design and Printing – Juliette Walker

Published by the Picnic School of Art 2022

Produced for the *Smartification of Everything* Conference and
Exhibition at the University of Ottawa, March 2022

Risograph printed at the UARK RISOlab in Fayetteville, Arkansas,
USA, February 2022

upon asymmetrical information access, ensuring who or what is relatively smart stays that way. This work invites experts and artists to consider the broad-reaching implications of smartification, with the hope to have contributed an experimental form from which future publications – and even publics – can be created, ideally to shed light on hyperlocal examples of information inequality that can nonetheless have global and historical import.

"The general thrust of any capitalistic logic of power is not that territories should be held back from capitalist development, but that they should be continuously opened up."

- David Harvey, *The New Imperialism*

Dockless shared electric scooters--smart scooters--seemed to arrive in Fayetteville, Arkansas, overnight, out of nowhere. A local news website, *The Fayetteville Flyer*, reported on June 27, 2019, that a company contacted the city and said that scooters would be dropped around town regardless of what anyone said. The assistant city attorney quoted in the article mentioned that a new state law must have been passed as a result of lobbying in Little Rock.[1] Regardless, the new law made it so the city couldn't ban the scooters outright. The scooters have been imposed, and superimposed on the city-- and they are a super imposition for me as a pedestrian in this place.

These scooters are often one of the first things that-- literally--cross my path each day. Whether one is overtaking me or impeding my progress on the sidewalk, they are impossible to ignore. There isn't much else in the built environment that is so transient and litter-like, yet replenished daily and designed to be so (aside from private automobiles, a topic for another time). My mind is boggled by these things that might just be too smart for me. Nonetheless, I consider why the scooters feel out of place: My neighbor doesn't leave their bike in the middle of the sidewalk or wherever they happen to stop riding it; they care about it getting banged up or stolen. If they don't lock it up properly in public, it could get stolen or cited by city officials. As a private citizen and not a venture-capital backed startup, my neighbor is not allowed to drop numerous bikes, which are also vending machines, around town and expect to turn a profit, let alone avoid legal action against them. What conditions make it possible for scooter companies to avoid the restrictions that would be imposed on my neighbor? I would suggest that the scooter startup executives should be required to live in the towns they leave their litter in if I were a lobbyist.

176 Devin Shepherd and Juliette Walker

The signs around town give me whiplash. One sign tells me motor vehicles aren't allowed on the bike path. But I just read on the City of Fayetteville's website that scooters are allowed on the bike path! The scooters have geofencing to slow them down on the University of Arkansas campus, but apparently not to prevent them from going down main street, where they are banned. Some small businesses have spray-painted signs on the pavement outside their doors reminding riders that scooters aren't allowed in this area; they look like guerilla advertisements--nothing like street signs or something I'd pay attention to while zipping by on a scooter. The scooters have signs on them, too, but not enough to understand how to use them without a smartphone or prior knowledge about how they work. The baseboards of the scooters say not to ride them on the sidewalk, but this is actually allowed, as long as it's not on the main street or otherwise in front of buildings with doors that open directly onto the sidewalk, which you can also learn by going to the city's webpage about this, although it's evidently not required reading by any means!

178 Devin Shepherd and Juliette Walker

In my inbox, I receive a newsletter from the University of
Arkansas every weekday. Once, the newsletter let me know that
a survey of scooter riders said the riders liked the scooters.[2]
It didn't say how many low-income people have taken advantage
of the programs that the scooter companies say will let them
use the scooters at a reduced cost if they have already
enrolled in state benefits of some kind and also shared this
information with the scooter company and have a bank account
and a phone and phone service that can be used to pay the
company with. It doesn't say how many people have not used
the scooters because they don't have a smartphone or a bank
account. The city website doesn't display this information
either, despite the state law specifying that the companies
must provide data to them--the only state law in the South that
does so, as of 2019.[3] I'm not a journalist, so I don't call
anyone to find out more. It seems like this information should
be publicly available, though--I'd go as far as to say this
data ought to be as unavoidable as the scooters cluttering the
sidewalks all over town.

The state collects taxes to fund public services, which can be
determined by the taxpayers through elected representatives.
Some say the government should be run like a business,
that markets should determine service levels. The bus in
Fayetteville, funded by taxes, is free at point of service, but
is about as reliable as the scooters are. This is because the
bus doesn't run on a strict timetable. But there is an app that
you can use to see if a bus is arriving soon. Does this make
the bus smart? If so, smart does not mean reliable! According
to the free-market, the bus and scooters are competitors for
my business, but as a potential customer, I'm not sure if I
am being competed for or against. At least I don't live in
a transit desert, which sounds natural, but just refers to a
place where the free-market won.

It stands to reason that being smart means knowing better. Smart is a relative term. This means that others must know less. So perhaps in order to remain smarter, it is easiest for the smartest to remain the smartest by preventing others from getting access to information, which maintains the relative scarcity of smartness. Who can afford to be smart in this way? I ask myself, rhetorically.

A scholar once said that artifacts have politics.[4] I imagine a scooter is self-aware, another kind of smart; so smart, it could speak. It claims it solves a problem (the way every previous technology has, without mentioning what new problems it might cause): there are big hills in Fayetteville; sometimes it's hot and humid. You could buy your own version of me to ride, but you can use me for a lot less up front, and treat me like a piece of garbage as a bonus. Sort of like Walmart-- you go to Walmart, right? Of course you do, there isn't really another option around here if we're both being honest. Don't worry, if I get destroyed, the venture capital money can pay for more of me--the landfill isn't going anywhere either! My plastic carcass will go there soon. I am very smart! The app you use to activate me on your phone is connected to a server that stores all kinds of data that gets collected while you ride me and use me for any purpose you like; this is for your own good and the good of the service. The computer also makes it possible for a person in a van to find out where I am. They drive around town and collect me and the other scooters when they need to be charged. You won't see your electricity bill go up because of me (it's not like you wouldn't be charging your phone either way), but you will pay enough to my owners to ensure that they will still make money. Don't think too much about it, or the environment, the wages of the gig worker who might have to venture onto private property to retrieve me, whether you should be wearing a helmet, or anything else really. All the thinking has been done for you! Also, please forget that buses and bicycles exist! I actually am better than cars, though; look it up! Cars could learn a thing or two from me. For instance, my built-in speed governor is actually sometimes activated.

When I first arrived at the University of Arkansas, pre-smart scooters, bikes had to be registered on campus or would be cited. Not anymore. "Maybe we have scooters to thank?" I ask myself as I realize all the bike racks are being occupied by so-called dockless scooters.

Smart can also mean pain, which is what happens to my brain when I think about smartification for too long.

Someone said recently that smartness just means surveillance.[5] Maybe so, but I feel like I'm the one who has to always be on alert--what was it Foucault said? Anyway, I have had to learn a lot about scooters against my will, and because I live where I do, I can say the same about Walmart. Walmart established the unmissability of low prices at the cost of shared space degradation: Walmarting. More like Walsmarting. It's a tale as old as the late 20th century: Don't want to shop at Walmart? Thoughts and prayers. Too late. The downtown (except the one in Bentonville, Arkansas) is a blight now because the prices were too high, the sights too myopic, the promised-but-not-guaranteed tax bump too good to pass up. No one wondered what would happen to a big box if it was unprofitable and abandoned after it put the competition out of business--that's a write off![6]

Vermont tried to keep Walmart out; Vermonters said no, Walmart said yes.

Walmart pioneered smartness. UPC codes, satellites, data centers--Walmart was an early adopter of such technologies.[7] Sam Walton could appear on a one-way satellite TV at any Walmart store to describe the numbers he was looking at (he couldn't see the employees).[8] Sam Walton always insisted it was about competition and giving the customer what the customer wants. Sam Walton considered technology to be a competitive advantage.[9] This is free-market smartness; dominate the market, become the only choice. Remarkably, one of the two sanctioned scooter companies operating in Fayetteville, Spin, recently got out of the "open permit" market in Germany because it lacked "sensible" regulations.[10] The company, owned by Ford, will now only operate in "limited vendor markets," like the duopoly enforced in Fayetteville. Compared to the United States, was Germany too open, too free? Walmart got out of Germany, too. Smart is as smart does.

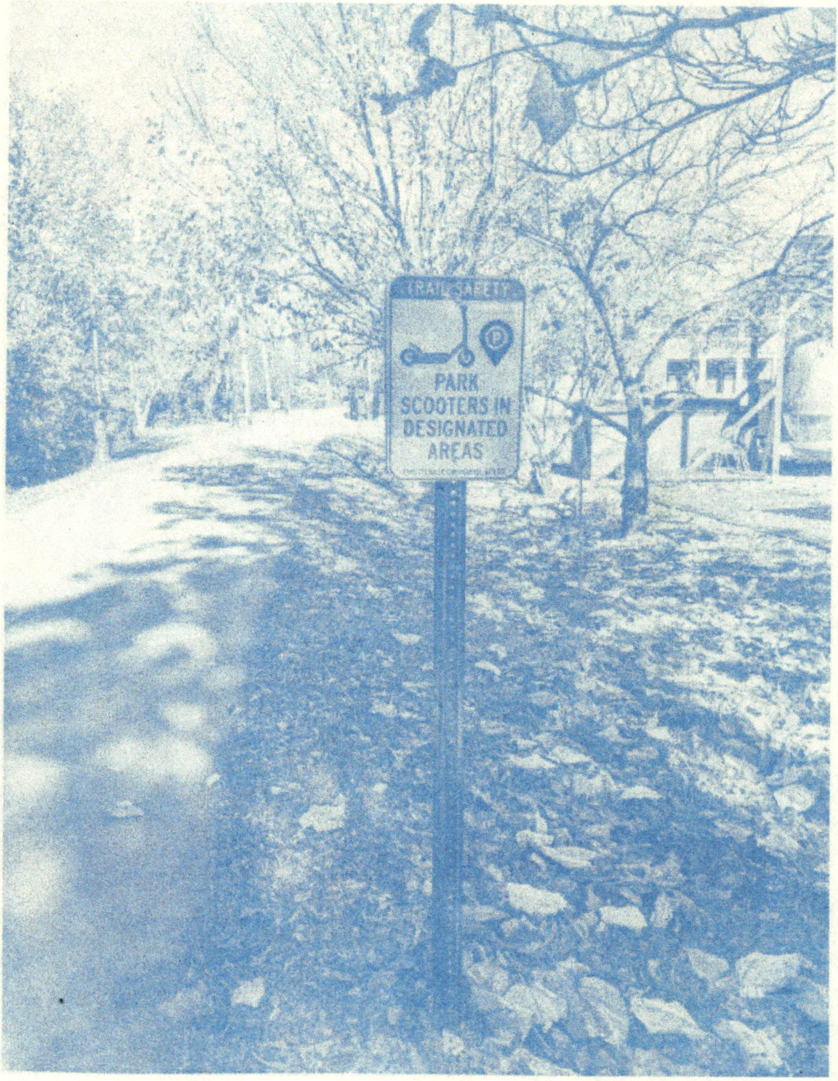
PARK
SCOOTERS IN
DESIGNATED
AREAS

So now I can't miss scooters in Fayetteville. But I've never been able to miss Walmart. In the past year, two brand new buildings funded with Walmart money have had groundbreakings on the University of Arkansas' campus--one of them is across the street from a Walmart! I wonder what I am missing, though. What would the University be like if Walmart didn't get to pick what got bought and paid for? Or Tyson. Or J.B. Hunt. The Big Three. I wonder what the scooter companies will someday buy to reinforce their being smart, not to mention magnanimous. I wonder if academe is the territory that is getting bought up now, as I search paywalled databases for articles about smart scooters. I notice that news articles make smartness seem inevitable; but I can't help but notice that advertisements for smartness are a lot easier to find than claims substantiating the benefits.

Wal*Smartification 187

Reference

Bajohr, Hannes. 2016. "Experimental Writing in Its Moment of Digital Technization: Post-Digital Literature and Print-on-Demand Publishing." In *Publishing as Artistic Practice*, edited by Annette Gilbert. Sternberg Press.

PART FOUR

Opening Up Smartness

17 Introduction to Opening Up Smartness

KELLY BRONSON

Although smartness has been imagined as, variously, an approach, an epistemology, a set of practices, and new technologies, the phenomenon is most closely associated with the city (Mattern 2021; Sadowski 2020). "Smart city" appears to be part of our everyday lexicon and indeed a part of many people's everyday experience given the dominance of thermostat-controlled buildings and sensing appliances. The chapters in part 4, "Opening Up Smartness," challenge the urban bias in popular and academic work on smartness. They open up and expose this biased thinking on smartness, and they additionally challenge the role that so-called smart tech plays in perpetuating stereotypical and harmful ideas of rural as distinct from urban society and from nature.

Martin Abbott in "FloodSmart: 'Equity in Action' or Equity Inaction?" suggests that the flip side of the dominance of the urban-centred imaginaries of smartness is a rural imaginary in which rural spaces are widely thought of, if they are thought of at all, as spaces devoid of cutting-edge technology and in need of development. Jean Hardy argues in "Rural Expertise and the Sewer" that rural spaces are considered largely in the context of attending to the impacts of natural disasters, at least in the political imagination. Yet Abbott questions what is so natural about natural disasters. And in thinking through this question, Abbott begins to unravel the seemingly hard and fast boundary between rural and urban, between environment and society (or artifice). Abbott shows how flooding is not so much a "natural" disaster as much as a product of human decision-making that has happened in very specific racial and political economic contexts. Floods only have disastrous consequences when the preconditions for flooding render people – and racialized people living in poverty – vulnerable. At the centre of Abbott's essay stands an algorithm which is proposed to help streamline flooding insurance provision and increase access. While an interesting proposition for a smart technology that addresses rural spaces, Abbott's analysis reveals that no algorithm for flood detection or insurance provision, no matter how sophisticated, can seek redress for or solve these Gordian knots of racial and economic politics as they intersect with land access and development in the United States. Said more directly: Abbott reveals a racial and economic bias within FEMA's algorithm for flood insurance provision – Risk Rating 2.0 – where the dataset and app presuppose an urban citizen who, regardless of race, class, and gender, has equal access to and lives in environmentally just places.

Using the case of AI-assisted traffic lights in Vienna, Austria, Poya Sepehr demonstrates how smartification processes could also "open up" opportunities for mobility justice. Jaywalking – the label given to people disregarding traffic light rules – can be regarded as rule-breaking or empowering, an act of navigating an unjust, car-dominated mobility system, with its "smart" traffic lights serving rather as devices of (social) control.

Similar to Abbott, S. Ashleigh Weeden in "Outsmarting Urbanism: Could Leveraging the 'Right to be Rural' Produce Alternative Futures?" challenges dominant notions that rural spaces are problems in need of technological solutions. Her essay takes up the question of what defines rural spaces and argues that "smartification agendas" (and their associated activities) tend to assume that rural spaces and the people who populate them need to be brought into the twenty-first century via smart technologies. Rural spaces are effectively empty (at least from a regulatory perspective) and thus places where smart technologies can be used to maximize the extraction of natural resources. The consequence of this dominant political stance towards the rural is a seeming inevitable urbanization. Weeden challenges this determinism and the dominant narrative of smartness qua urbanization by centring rural spaces as embedded with features (e.g., greater diversity than cities) that actually provide solutions to social and environmental problems. The result? A "toolkit" for challenging smartification agendas and advocating for the "right to rurality."

Jean Hardy raises a similar political agenda in his chapter "Rural Expertise and the Sewer" by highlighting the crucial role of experiential and local knowledge held by rural citizens when compared to so-called smart sewage infrastructure. Using the city of Houghton's attempted implementation of GIS-informed stormwater and wastewater management as a case, Hardy explores how the digital systems developed under a grant meant to enable smart development ultimately failed to gain traction given the specific knowledge environment and context of Houghton. Indeed, this smart infrastructure project confronted the kind of diversity that Weeden discusses, thus corroborating her point that smartification agendas often mistakenly assume a standardized rural space to the detriment of those spaces and the smartification projects. In Hardy's case study, the smartification agenda of the sewer and wastewater asset management grants was to "modernize" the maintenance of public infrastructure through the creation and central interoperable storage of digitized data about sewer and wastewater infrastructure. The agenda promised to transform how the municipality maintained and tracked infrastructure and indeed to democratize infrastructure via access to important data. However, the communities did not have the funds to use the big data and found that human expertise – that of local municipal employees who knew the small-scale sewer systems well – functioned better than the so-called smart one. Thus, In Hardy's case, we not only see the negative implications of the hubris of Weeden's smartification (and urbanizing) agendas, but we also confront the limitations in assuming that smartness is embedded in technology and not human experience.

References

Mattern, Shannon. 2021. *A City Is Not a Computer: Other Urban Intelligences.* Vol. 2 of *Places Books.* Princeton University Press. https://doi.org/10.1515/9780691226750.

Sadowski, Jathan. 2020. *Too Smart: How Digital Capitalism Is Extracting Data, Controlling Our Lives, and Taking Over the World.* MIT Press. https://doi.org/10.7551/mitpress/12240.001.0001.

18 FloodSmart: "Equity in Action" or Equity Inaction?

MARTIN ABBOTT

On Being Flood Smart

In May 2021, LaToya Butler-Jones opened her webinar presentation, "FloodSmart," with a poll: "What is America's number one natural disaster? How many states suffer from flooding?" (USACE 2021). Butler-Jones, a marketing liaison at the Federal Insurance and Mitigation Administration (FIMA), highlighted the limited public understanding of flood insurance. A Federal Emergency Management Agency (FEMA) subagency, FIMA manages the US government's National Flood Insurance Program (NFIP). With this public knowledge deficit as its starting point, Butler-Jones's presentation aimed to "spread the word on how to be flood smart" (USACE 2021).

FloodSmart is a long-running FEMA outreach and marketing campaign to educate the public about the NFIP (Kousky 2018). The NFIP provides flood insurance to property owners and renters in over twenty thousand participating communities. In many of these participating communities, subsidized flood insurance underpins local housing markets. Without it, critics claim, coastal communities at risk of flooding, such as New Orleans, could struggle for survival if housing prices declined due to flood insurance becoming too expensive, too quickly (CfSFI 2016). The campaign's central node is FloodSmart.gov, a website established to "tell the story of flood risk and drive policy sales" (FEMA 2022a). To promote the financial and psychological benefits of insuring against flood loss, FloodSmart.gov showcases the virtues of protection and recovery to prospective customers online. Both the practice of being flood smart and FloodSmart.gov offer insight into a raft of tools that sell the idea of flood insurance as an appropriate fix to flooding disasters.

Arguably more important to being flood smart is the story FloodSmart.gov does not tell. When the NFIP was established in 1968, it was anticipated that the program would protect the nation's largely white homeowners from financial loss (Elliott 2021). Since then, little effort has been made to reform the program's racialized foundations. The representation of the average American home on FloodSmart.gov, for example, reflects an almost unrecognizable peri-urban ideal rather than the cosmopolitan urban fabrics of the USA where most Americans live today (USCB 2023).[1] Furthermore, the website makes no mention of the uneven distribution of flood risk that burdens some communities and not others.

In this chapter, I show how smartness is mobilized as both a rhetorical practice to market and sell flood insurance and a way of knowing flood risk. From FEMA's standpoint, a homeowner or renter is flood smart when they have been educated to know that they should buy flood insurance. As emerging smart technologies reshape the NFIP, however, scrutiny of being flood smart calls into question how and why flooding affects some people more than others. I argue that the label "smart" acts to conceal rather than confront a history of inequity in FEMA's flood protection and recovery programs. I also argue that the label of "smart" and technologies associated with it perpetuate a long history of environmental inequity and racial injustice. To do this, I draw on ten months of ethnographic field research in New Orleans. I also draw on digital ethnographic research undertaken during the pandemic. Over fifty interviews with experts and professionals in the hazard mitigation arena also provided important background for the chapter. Focused on New Orleans, I interrogate FloodSmart and "Risk Rating 2.0 – Equity in Action" (RR2.0), an emerging smart flood technology that is redefining the NFIP. In putting these ostensibly "smart" components of the NFIP into conversation, I ask: what does it mean to be flood smart?

Making Space for the Social Study of Smart Technology in the City

This chapter draws on scholarship situated at the intersection of urban studies and science and technology studies (STS). In these interdisciplinary fields, the concept of "smartness" is ambiguous (Hommels 2020). In the city, what is and isn't considered "smart" is difficult to distinguish

---

[1] According to the United States 2020 Census (2023), four in five Americans or 265 million people live in urban areas.

because information and communication technologies (ICT) continue to proliferate. This section elaborates the theoretical contours of both fields to push for the social study of smart technologies in the city.

In urban studies, the concept of the "smart city" is often the locus of inquiry when investigating the application of science and technology in the city. Scholars emphasize the fuzziness of the concept and absence of standard definitions (Albino et al. 2015). Others note that the concept is employed inconsistently across different sectors (Greenfield 2013; Joss et al. 2019). Overall, two key attributes are observed (Glasmeier and Christopherson 2015).

First, technological progress is said to enhance the management of urban systems and increase wealth creation. Second, smartness is a futuristic urban creation that imagines new realities. Both findings attest that smartness is not only relevant to ICT hardware but also the softer social and relational dimensions of urban society. In this guise, smartness is a tool to organize, manage, and govern the fabric of the city. In the event of flooding, the smart city's incongruous sibling – the smart disaster – similarly promotes the utility of state-of-the-art technofixes at the expense of socially grounded alternatives (Boukerche and Coutinho 2018). Critical readers of the smart city highlight the dearth of detailed empirical research (Kitchin 2015). Throughout the chapter, I use "smart technology" to heighten engagement with society and spaces in, beyond, and between the city as well as STS scholarship.

In STS, scholars show how smart technologies are shaped by the extractive business practices of corporations and how these technologies shape urban society. To sell software and services to governments, corporations like IBM, Cisco, and Google enact powerful imaginaries that portray data analytics and "smart" infrastructure as the saviour of the city while seldom delivering on their promised outcomes (Halpern et al. 2013). Smartness in this context is presented as a "panacea" for any number of "demographic, economic, and ecological challenges" (Farías and Blok 2016, 555). Smartness, therefore, leaves little room for other imagined urban futures to flourish (Miller 2020). The critical and reflexive approach of STS in the city also foregrounds how the knowledge and values that produce off-the-shelf technologies are always situated and subsequently rarely function seamlessly out of context. STS approaches challenge how knowing flood risk uniformly across the USA, vis-à-vis knowing that flood risks differ in New Orleans or New York City, changes our conception of what smartness is and how environment-society relations might be otherwise. Ultimately, in contesting the multiple ways of knowing cities as situated places, STS analysts interrogate the dynamic and multidirectional relationships between social, environmental, and technoscientific relations (Mattern 2017).

The Cost of Flooding

Upon visiting the FloodSmart website, prospective flood insurance customers are greeted with a stock photo of a smiling Black and Brown couple seated in front of a laptop computer. The image conveys the impression that the couple are becoming flood smart. In a nondescript domestic space, the couple are presumably weighing the costs and benefits of a flood insurance policy. To the left of the photo, a blue text box explains that "the NFIP offers flood insurance to help you protect the life you've built and recover more quickly after a flood" (FEMA, n.d.). Below the photo, stylized calamity icons and text in austere colours are superimposed on a white background. This menu of subpages includes the "Cost of Flooding," "What's Covered?" "Risk Rating 2.0: Equity in Action," and "Follow Your Instincts This Hurricane Season." On these subpages, FEMA lays out the virtues of flood insurance. The website bristles with not only positivist and rational data points to quantify flood risk but also interpretive statements about recovering emotionally in the aftermath of a flooding disaster.

On the subpage dedicated to the "Cost of Flooding," an interactive tool illustrates FEMA's approach. The tool declares, "Just 1 inch of water can cause $25,000 of damage to your home" (FEMA 2022b). The tool, designed around a caricature of a living room, showcases a stylish mid-century modern couch. Hanging on the wall above the couch is a framed picture of an idyllic peri-urban home. This "average" American home, defined by FEMA for the purposes of estimating the financial cost of flooding, is a detached, slab-on-grade, 2,500 square-foot, single-story dwelling. On one side of the house, cows roam in a timber-fenced field. On the other side, a creek flows past the house. Beyond the creek, a fox roams freely among the pines. Parked on the grass in front of the house is a car with a canoe strapped to the roof. Lapping at the legs of the couch in this idyllic scene, however, is an inch of floodwater. The tool's interactivity is largely attributable to a slider that changes the height of the floodwater. If one foot of water flooded this 2,500 square-foot family home, the cost of repairs would be US$72,162. The slider takes the floodwaters no higher than four feet and indicates repairs at this level of inundation would cost US$103,355. At this point, having completely submerged the couch, the floodwater laps at the picture frame. In addition to FEMA's standard 2,500 square-foot home, the interactive tool offers customers the choice of either a "small" 1,000 square-foot single-story home or a "large" 5,000 square-foot double-story home. The tool asks, "Which home is most like yours?"[2] Regardless of the cost and benefit of buying flood insurance, the tool emphasizes that flooding is also an emotionally devastating event.

The FloodSmart website epitomizes FEMA's efforts to turn victims into beneficiaries of disaster. One of the primary objectives of the NFIP's policy's architects was to "transform the flood victim into an insurance customer" (Elliott 2021, 44). The goal is to "turn the post-event, ad hoc nature of compensating the victims of acts of God into a consistent, nationally standardized response to flood loss" (Elliott 2021, 44). This is achieved through the prepayment of an insurance premium, which then gives homeowners and renters the right to make a claim in the event of a flood. This process is what the FIMA liaison would describe as "being flood smart." If you are smart, you will buy flood insurance no matter where you live (FEMA 2022c). By transforming a flood into an insurable event, the NFIP has "the effect of normalising disaster" (Elliott 2021, 45). By normalizing disaster, the program also normalizes inequity. In the following section, I unpack an emerging smart technology that seeks to enhance equity in FEMA's disaster protection and recovery programs.

"Equity in Action" or Equity Inaction?

FEMA's story of protection and recovery fails to mention how urban flooding caused by heavy rainfall over the built environment remains difficult to digitally model and map. In New Orleans, the difficulty lies at the intersection of a dynamic array of social, environmental, and technoscientific relationships. To overcome this, FEMA has introduced a transformational smart technology. Described by FEMA as "Risk Rating 2.0 – Equity in Action," the agency claims this new methodology represents a seismic shift in how flood risk and the price of insurance are determined. The algorithm that underlies RR2.0 is the first update to the NFIP's pricing methodology in fifty years. FEMA (2022d) contends that the new algorithm underlying this smart technology, unlike the old methodology, will "deliver rates in a more modern, individualized, and equitable way." Guided by cost-benefit analysis, however, RR2.0 is wholly focused on financial equity rather than environmental equity or other social vulnerabilities.

---

[2] With four feet of water in the living room, the cost of repairs would be US$43,400 for the smaller home and US$203,280 for the larger home. These alternatives demonstrate that the cost of flooding is somewhat less for the former and considerably more for the latter.

Across the USA and especially in New Orleans, low-income residents and people of colour live in high flood-risk areas and are disproportionately affected by flooding (Wing et al. 2022). In the City of New Orleans's (2015) *Urban Resilience Plan*, a map shows how the distribution of flood risk is a burden to some and not others. Titled "Disproportionate Risk," the map shows that communities of colour bear higher risk burdens in the areas that are at greatest risk of flooding. While the inequity of flood risk is a contemporary concern, it is deeply embedded in historical planning practices that racially segregated urban America (Rothstein 2017; Taylor 2019). For much of the twentieth century, redlining, among other modes of calculation, pushed people of colour and low-income residents in New Orleans to live in areas that were either on the periphery or at risk of flooding. Until the Civil Rights Act of 1968 and its fair housing provisions, the practice of redlining enacted by successive US governments systematically excluded people of colour from the housing market by making it difficult for them to access mortgages and build homes in desirable neighbourhoods, such as those that did not regularly flood. Other scholars show how these racist practices endure (Anderson 2022). This is why FEMA flood maps "bear a striking resemblance to 1930s redlining maps" (Katz 2021). Disproportionate impacts are also baked into FEMA's disaster relief.

Equally important to being flood smart is the story of recovery. In the aftermath of recent flooding disasters, environmental justice scholars have highlighted the inequitable impact of FEMA's relief programs. FEMA's programs favour white communities and exacerbate wealth inequality. For example, after Hurricane Katrina, the "formula used to allocate grants to homeowners through the Road Home program – 'the single largest housing recovery program in U.S. history'" – discriminated against African Americans and instigated a civil rights case (Bullard and Wright 2012). Howell and Elliot quantify this uneven distribution of relief, showing that "as local hazard damages increase, so too does wealth inequality, especially along the lines of race, education, and homeownership. The more funds areas receive from FEMA, the more this wealth inequality increases" (2019). Compiled data from 1999 to 2013 shows that white areas grew, on average, US$126,000 richer in counties extensively damaged by natural hazards. By contrast, in the same circumstances, Blacks and Latinos lost US$27,000 and US$29,000, respectively.[3]

---

3 Step further back in time to the Great Flood of 1927 when the Caernarvon levee channelling the Mississippi River was blown up to save New Orleans. Historian Richard Mizelle Jr. details how unjust "racial customs and norms were nonetheless maintained during this human-made disaster" (2014, 45).

"The pattern of recovery," to summarize, "has been that money follows money; money follows power; and money follows Whites" (Bullard in NASEM 2021).

Late in 2021, I attended another webinar titled "Risk Rating 2.0: What Floodplain Managers Need to Know," organized by the Association of State Floodplain Managers in concert with FEMA. The Association of State Floodplain Managers is the peak body representing the US floodplain management community and consists of public officials and private members. In their introduction, just like the FIMA marketing liaison, the veteran communications consultant promised attendees that they were in for a "real learning experience." Looking over the many questions asked by attendees, I observed that few people understood RR2.0. This uncertainty was consistent with other flood insurance webinars that I have attended. But FEMA can neither explain how the myriad risk variables are weighted nor how the algorithm works because it is a proprietary system that uses data from private sources.[4]

FEMA points out they are "leveraging improved technologies" to increase the number of risk variables used in determining the price of flood insurance. Given the equity focus of the new pricing methodology, one could imagine that environmental and socioeconomic factors would be included in the algorithm. The algorithm variables listed in FEMA's (2021) extensive *RR2.0 Methodology and Data Sources* report, otherwise known as the *Milliman Report*, however, are exclusively focused on geographic factors, such as the distance of a property to a river or the ocean, and property characteristics like the height of the first floor above the ground. Moreover, catchwords like "equity," "climate change," and "environment" are not to be found anywhere in this document.[5] On the FloodSmart website, too, a search for "equity" shows the term is solely associated with RR2.0. Environmental equity and racial justice, among other social vulnerabilities, do not appear to be factored into what FEMA describe as their engine of equity.

The Value(s) of Flood Protection and Recovery

Like past NFIP reforms, a major concern is that as the influence of RR2.0 grows, the price of individual insurance premiums will increase by thousands of dollars. Critics also fear that as prices rise, the number

---

[4] At the time of writing, the question of how the algorithm works has become the focus of a legal challenge.
[5] "Environment" is used twice in the document in reference to "regulatory environments" and an "Environmental Protection Agency link."

of insurance policy holders will decline nationally, which could threaten the viability of a program mired in more than twenty billion dollars of congressional debt. The concern is that this reform could end publicly subsidized flood insurance. In coastal communities across the USA and in New Orleans in particular, subsidized flood insurance is the bedrock of local housing markets. In New Orleans, this may mean low-income residents and people of colour who live on the floodplain could be at a disadvantage because insurance premiums will rise substantially to reflect the actual risk of flooding. Can the engines of algorithmic technologies like RR2.0 be finetuned and remade "FloodSmart" to tell the story of environmental equity and racial justice?

The contradiction inherent in FEMA's emerging engine of equity draws attention to the reproduction of inequities that have been baked into the NFIP. In normalizing disaster through insurance, the NFIP was established as a "safety net" that would "catch everyone with a policy, which is to say, the largely white homeowners for whom previous decades of housing policy had facilitated access to property ownership" (Elliott 2021, 45). Since the NFIP's establishment, therefore, it has also normalized inequity in places like New Orleans because federal policies at this time imagined the "typical urban citizen" to be a "white, cis, able-bodied, middle class, heterosexual man" (Kern 2020, 49). In New Orleans, this imagined urban citizen reflects only a small minority of the majority Black city.

The positionality of which urban citizens FEMA renders flood smart raises important questions about the ongoing smartification of the NFIP. From FEMA's standpoint, on the one hand, encouraging citizens to become flood insurance customers is smart. With more policyholders, the risk pool expands, and revenues grow. Coupled with a new individualized pricing methodology that is determined by an engine of (in) equity, FEMA imagines that all urban citizens, regardless of race, class, and gender, are equal and live in just places. As I have shown above, FEMA's universal imaginary of the urban citizen neither reflects New Orleans's population nor the demography of the USA. On the other hand, FEMA's mobilization of smartness as a way of knowing flood risk in order to market and sell flood insurance suggests that they are trying to save those citizens who have not already purchased flood insurance. This deficit identifies them as prospective customers. This then renders those who buy flood insurance as smart and leaves those who do not or cannot buy insurance looking stupid. In New Orleans, the positionality of who is rendered stupid is deeply racialized because of the inequities baked into FEMA's flood protection and recovery programs and the spatiality of flood risk that burdens communities of colour more than others. Instead of rendering some citizens smart and others stupid, the smartification of flood insurance could acknowledge the environmental inequities and racial injustice and work to reform them.

Conclusion: Rethinking Flood Smarts

This chapter has shown how smart technologies are mobilized by FEMA in the NFIP. FEMA employs smartness as both a rhetorical practice to market and sell flood insurance and a way of informing the public about flood risk. Two smart technologies were analysed. The first was FloodSmart, a long-running marketing campaign; the second was "Risk Rating 2.0 – Equity in Action," an emerging smart flood technology. In this context, FEMA imagines that urban citizens are flood smart when they have been educated to know that they should buy flood insurance. Consequently, this may leave many looking flood stupid.

My analysis of what it means to be flood smart shows how the same mid-twentieth-century policy architecture that established the NFIP continues to shape outcomes through RR2.0. This new smart flood technology, in particular, risks perpetuating the same inequities and injustices that are the hallmark of FEMA's flood protection and recovery programs. The "smart" label serves to legitimize continued inequity by projecting an overblown image of improvement that is ultimately hollow. My analysis also shows how FloodSmart rhetoric, as well as the appeal of algorithmic technologies, serves to obscure rather than communicate a more nuanced story about the NFIP.

As climate change impacts grow, we must pay careful attention to how stories of smart technologies are told about flood protection, recovery, and resilient futures. To do this, a truly smart and equitable technology should take into consideration not only economic concerns but also social and environmental factors.

References

Albino, Vito, Umberto Berardi, and Rosa Maria Dangelico. 2015. "Smart Cities: Definitions, Dimensions, Performance, and Initiatives." *Journal of Urban Technology* 22 (1): 3–21. https://doi.org/10.1080/10630732.2014.942092.

Anderson, Elijah. 2022. *Black in White Space: The Enduring Impact of Color in Everyday Life.* The University of Chicago Press. https://doi.org/10.7208/chicago/9780226815176.001.0001.

Boukerche, Azzedine, and Rodolfo Coutinho. 2018. "Smart Disaster Detection and Response System for Smart Cities." In *2018 IEEE Symposium on*

Computers and Communications. IEEE. https://doi.org/10.1109/ISCC.2018.8538356.

Bullard, Robert, and Beverly Wright. 2012. *Race, Place, and Environmental Justice After Hurricane Katrina: Struggles to Reclaim, Rebuild, and Revitalize New Orleans and the Gulf Coast.* Westview Press.

City of New Orleans. 2015. "Resilient New Orleans: Strategic Actions to Shape Our Future City." https://resilientcitiesnetwork.org/downloadable_resources/Network/New-Orleans-Resilience-Strategy-English.pdf.

Coalition for Sustainable Flood Insurance (CfSFI). 2016. *Making the NFIP Work for Taxpayers and Policy Holders: Policy Affordability and Program Stability.* https://csfi.info/wp-content/uploads/2017/08/CSFI-Making-the-NFIP-Work-for-Taxpayers-and-Policyholders-Policy-Affordability-and-Program-Stability-.pdf.

Elliott, Rebecca. 2021. *Underwater: Loss, Flood Insurance, and the Moral Economy of Climate Change in the United States.* Columbia University Press. https://doi.org/10.7312/elli19026.

Farías, Ignacio, and Anders Blok. 2016. "STS in the City." In *The Handbook of Science and Technology Studies*, 4th ed., edited by Ulrike Felt, Rayvon Fouché, Clark Miller, and Laurel Smith-Doerr. MIT Press.

FEMA. 2021. "Risk Rating 2.0 Methodology and Data Sources." www.fema.gov/sites/default/files/documents/fema_risk-rating-2.0-methodology-data-sources_4-21.pdf.

— 2022a. "FloodSmart: Resource Library." https://www.fema.gov/flood-insurance/outreach-resources.

— 2022b. "FloodSmart: The Cost of Flooding." https://web.archive.org/web/20240217034300/www.floodsmart.gov/cost-flooding.

— 2022c. "FloodSmart: Hurricane Season." www.floodsmart.gov/quiz/Q1/no.

— 2022d. "FEMA Fact Sheet – Understanding Risk Rating 2.0: Equity in Action." Accessed July 20, 2023. https://web.archive.org/web/20230822125717/https://agents.floodsmart.gov/sites/default/files/fema-Risk-Rating-2.0-Fact-Sheet-2022.pdf.

— n.d. "FloodSmart: The National Flood Insurance Program." Accessed July 20, 2023. www.floodsmart.gov/.

Glasmeier, Amy, and Susan Christopherson. 2015. "Thinking About Smart Cities." *Cambridge Journal of Regions, Economy and Society* 8 (1): 3–12. https://doi.org/10.1093/cjres/rsu034.

Greenfield, Adam. 2013. *Against the Smart City.* Do Projects.

Halpern, Orit, Jesse LeCavalier, Nerea Calvillo, and Wolfgang Pietsch. 2013. "Test-Bed Urbanism." *Public Culture* 25 (2): 272–306. https://doi.org/10.1215/08992363-2020602.

Hommels, Anique. 2020. "STS and the City: Techno-Politics, Obduracy and Globalisation." *Science as Culture* 29 (3): 410–16. https://doi.org/10.1080/09505431.2019.1710740.

Howell, Junia, and James Elliott. 2019. "Damages Done: The Longitudinal Impacts of Natural Hazards on Wealth Inequality in the United States." *Social Problems* 66 (3): 448–67. https://doi.org/10.1093/socpro/spy016.

Joss, Simon, Frans Sengers, Daan Schraven, Federico Caprotti, and Youri Dayot. 2019. "The Smart City as Global Discourse: Storylines and Critical Junctures across 27 Cities." *Journal of Urban Technology* 26 (1): 3–34. https://doi.org/10.1080/10630732.2018.1558387.

Katz, L. 2021. "A Racist Past, a Flooded Future: Formerly Redlined Areas Have $107 Billion Worth of Homes Facing High Flood Risk – 25% More Than Non-Redlined Areas." *Redfin News*, March 14. www.redfin.com/news/redlining-flood-risk/.

Kern, Leslie. 2020. *Feminist City: Claiming Space in a Man-Made World.* Verso.

Kitchin, Rob. 2015. "Making Sense of Smart Cities: Addressing Present Shortcomings." *Cambridge Journal of Regions, Economy and Society* 8 (1): 131–6. https://doi.org/10.1093/cjres/rsu027.

Kousky, Caroline. 2018. "Financing Flood Losses: A Discussion of the National Flood Insurance Program." *Risk Management and Insurance Review* 21 (1): 11–32. https://doi.org/10.1111/rmir.12090.

Mattern, Sharon. 2017. *Code and Clay, Data and Dirt: Five Thousand Years of Urban Media.* University of Minnesota Press. https://doi.org/10.5749/minnesota/9781517902438.001.0001.

Miller, Thaddeus. 2020. "Imaginaries of Sustainability: The Techno-Politics of Smart Cities." *Science as Culture* 29 (3): 365–87. https://doi.org/10.1080/09505431.2019.1705273.

Mizelle, Richard, Jr. 2014. *Backwater Blues: The Mississippi Flood of 1927 in the African American Imagination.* University of Minnesota Press. https://doi.org/10.5749/minnesota/9780816679256.001.0001.

National Academies of Sciences, Engineering, and Medicine (NASEM). 2021. *Perspectives on Climate and Environmental Justice on the U.S. Gulf Coast: Proceedings of a Webinar – in Brief.* The National Academies Press. https://nap.nationalacademies.org/catalog/26348/perspectives-on-climate-and-environmental-justice-on-the-us-gulf-coast.

Rothstein, Richard. 2017. *The Color of Law: A Forgotten History of How Our Government Segregated America.* Liveright.

Taylor, Keeanga-Yamahtta. 2019. *Race for Profit: How Banks and the Real Estate Industry Undermined Black Homeownership.* UNC Press. https://doi.org/10.5149/northcarolina/9781469653662.001.0001.

United States Army Corps of Engineers (USACE). 2021. "FloodSmart: A Guide to Flood Insurance." https://usace.contentdm.oclc.org/digital/collection/p16021coll2/id/6841.

United States Census Bureau (USCB). 2023. "2020 Census: Urban Areas Facts." June 29. www.census.gov/programs-surveys/geography/guidance/geo-areas/urban-rural/2020-ua-facts.html.

Wing, Oliver, William Lehman, Paul Bates, Christopher Sampson, Niall Quinn, Andrew Smith, Jeffrey Neal, Jeremy Porter, and Carolyn Kousky. 2022. "Inequitable Patterns of US Flood Risk in the Anthropocene." *Nature Climate Change* 12 (2): 156–62. https://doi.org/10.1038/s41558-021-01265-6.

19 Should I Stay, or Should I Go? Questioning the "Smartness" of Intelligent Pedestrian Traffic Lights in Vienna

POUYA SEPEHR

Introduction

The modern city has long been conceptualized as the confluence of technological innovation and organic development – a fusion described by Shannon Mattern as the "merger of the technological and the organic" (2021, 59). Various metaphors – machine, organism, network, system, or a system of systems, and more recently, as intelligent and smart – reflect the evolving understanding of urban environments. These descriptors not only characterize the city but also prescribe methodologies for managing its complex dynamics. In the era of the contemporary smart city, these methodologies are increasingly informed by digital technologies, which promise new capabilities under the guise of efficiency and improved governance. However, as Jathan Sadowski (2020) points out, the practical implementation of these smart technologies often translates into extensive data collection, enhanced network connectivity, and increased control.

This chapter aims to dissect the implications of such enhanced control, particularly through software automation, as cities transform into interconnected, seamless, and technologically advanced systems – a phenomenon referred to as the "smartification" of everything in this edited volume. Specifically, the city of Vienna, a proponent of the smart city agenda, has embarked on an initiative to automate pedestrian signalling at crosswalks using smart surveillance cameras and algorithmic computation as showcased more extensively in Sepehr (2024). Launched in 2020, this intelligent pedestrian traffic light system detects the intent of pedestrians to cross and alters the signal accordingly, aiming to mitigate jaywalking and streamline urban traffic flow. Although localized, this initiative underscores the broader impact of the smart city paradigm, revealing how urban spaces are reimagined and reconfigured through the lens of computational urbanism (Tironi and Valderrama 2018).

Urban anthropologist Shannon Mattern (2021) critiques the prevalent metaphor of the city-as-computer, arguing against its simplistic view of urban complexities as merely programmable entities subject to logical order (Mattern 2021, 62). In this context, smart urbanism, governed by the logic of algorithmic epistemologies, introduces a new form of order and control, challenging the tenets of democratic urban governance. Mattern thus urges scholars to explore the roles that specific technologies and initiatives play within the broader framework of urban governance, particularly focusing on how these technological assemblages influence and shape urban life (2021, 49). The intelligent pedestrian traffic light in Vienna exemplifies such a technological intervention, providing a pivotal case study for examining the intersection of technology, design, and urban governance in the smart city landscape.

My contribution in this edited volume on AI-assisted traffic lights in Vienna integrates into the third theme: "Opening Up Smartness." This theme encourages a deep dive into the layers of smart technologies, examining how they intertwine with and impact the social, political, and environmental fabrics of urban life. My analysis explores the dual role of traffic lights as both tools for urban efficiency and instruments of social control, providing a concrete case of how smart technologies shape public spaces and behaviours.

A Smart Solution to a Dumb Problem

The global allure of "smart" solutions has led to the "smartification" of virtually everything in our daily lives over the past two decades. However, this trend has raised concerns about emerging forms of power and control. As Sadowski (2020) illustrates, smart technologies inherently serve the interests of a select few who possess the means to control them. These technologies, while promising efficiency, often address what are essentially "dumb" problems, diverting attention from more pressing issues. They produce both intended and unintended consequences, with effects that are both known and unknown. The role of critique is to challenge the notion that these "smart" imperatives are absolute truths and to expose the politically driven intentions behind them. The challenge lies in redefining our understanding of "smartness" (Sadowski and Bendor 2019) and questioning the means and ends of smartification. In the context of smart cities, Sadowski (2020) contends that the political economy of smartness primarily serves the interests of a few elitist technopowers, often bypassing accountability and ethical considerations associated with smartness. This has led to the emergence of IT expertise as new "urban enablers" (Seghrouchni et al. 2016), tasked with transforming

Figure 19.1. The pushbutton box, which is called a *Bettlerampel* or begging traffic light in Vienna.

urbanism into techno-solutionism. The case of intelligent pedestrian traffic lights in Vienna provides a clear example.

There are around 1,300 pedestrian traffic lights in use in Vienna, of which 200 are equipped with a pushbutton box (fig. 19.1). These are traffic lights for pedestrians, who must push a button for the adjacent signal to turn red for the cars and let the pedestrians cross – in German these traffic lights are called *Bettlerampel*, which means begging traffic light. They are only in operation on busy roads where there is little pedestrian traffic, thus not many people know that they need to activate them in order to change the signal. The issue with pushbutton boxes, as a traffic engineer from the City of Vienna explained to me, is that "they lead to misunderstandings and confusion since pedestrians expect the traffic signal to turn for them automatically – that [is] how every other pedestrian traffic light works in Vienna." To address this rather trivial issue, the city's Traffic Light Department (MA 33) decided to hire the research group for computer vision and computer graphics surveillance technologies from the Technical University of Graz (TUG) to design a system that can automatically recognize when pedestrians intend to cross a junction so that the traffic light will turn red without the need to be activated by the pedestrian (Ertler et al. 2018).

Figure 19.2. The similarity between the acoustic signal system and the pushbutton box.

To understand the issue better, the lead research scientist from TUG explained to me that "Firstly, pedestrians may not realize that they need to activate the traffic light using the pushbutton box, which could cause some delay." This confusion is primarily due to the box's resemblance to the acoustic signal system designed for visually impaired individuals (fig. 19.2). The scientist continued: "Even when the pedestrian pushes the button, he or she may soon cross the street as soon as there are no cars [jaywalking]." However, there is another effect here. "After the person [the jaywalker] crosses the street, then the traffic light turns red and makes the cars to stop despite there is no pedestrian." Basically, the problem is unnecessary waiting time mostly for the cars but also for the pedestrians who do not know about the pushbutton box or who do not jaywalk. Hence, the goal is to get rid of the pushbutton box.

The research team from TUG developed learning algorithms that recognize pedestrians' intention to cross the street in less than four seconds. The system offers three main functions: automatic green phase initiation, signal change cancellation if the pedestrian crosses before the signal turns green, and automatic extension of the green phase for larger groups. Despite these features, the system faced challenges during its trial phase in Vienna.

The design process failed to consider the needs of visually impaired people, who rely on the box's speaker to navigate their crossing. Additionally, a safety margin is still required before the traffic signal is allowed to change, rendering the four-second detection meaningless in practice. This case illustrates how the allure of smart solutions can sometimes overshadow the simplicity of the problems they aim to solve. The intelligent traffic light project, while technologically advanced, essentially addressed a confusion issue. Moreover, it failed to consider the needs of all users, particularly the visually impaired. This highlights the importance of inclusive design and the need for public-expert comments during the feasibility study phase. The smartification of cities should not only focus on technological advancements but also on addressing the needs of all citizens in a democratic and inclusive manner (Sepehr and Felt 2023).

This case, however, is far from closed. It opens a window to broader sociotechnical issues, particularly the transformation of streets through control mechanisms like traffic lights. As we delve deeper into the narrative, we uncover more dimensions that highlight the intricate interplay between society and technology. The story of smartification continues to unfold, reminding us that even the most trivial issues can illuminate profound insights into our technologically mediated urban existence, as argued by Woolgar and Cooper (1999). The justifications surrounding the intelligent traffic lights – their purpose, their impact, and their role in the city – play a crucial role in the ways in which smartification is normalized. In what follows, I will take a more historical view of the emergence of traffic lights as an object of control and order in the city and the profound impact they have had on public space.

Questioning the Normalization of Smart Traffic Lights

The introduction of intelligent pedestrian traffic lights in Vienna as part of its smart city initiatives serves as a pertinent example of how technical artefacts become normalized within urban landscapes. Drawing on Langdon Winner's (1980) theory that technical artefacts are inherently political, these traffic lights manifest Vienna's policies and aspirations towards smart urbanism. However, as Joerges (1999) argues, the politics of artefacts are relational and depend on the interactions these artefacts have with other entities. In this context, intelligent traffic lights are not merely tools for automating pedestrian crosswalk signalling; they are instruments for reshaping urban spaces and redefining the city's relationship with its citizens. To fully grasp this dynamic, it is crucial to explore the evolution of traffic lights as objects of control that have historically reshaped streets and urban interactions.

Undeniably, traffic lights are among the most successful inventions of our time. Few inventions have achieved the universal status of the traffic light signalling mechanism (McShane 1999). They symbolize a potential global law and order mechanism (Zeno-Zencovich 2016). The invention and universalization of traffic lights tell a fascinating story, narrating the transformation of streets from predominantly pedestrian spaces to thoroughfares for cars. Peter Norton (2011) in *Fighting Traffic* details how American streets changed dramatically during the early twentieth century. As the number of automobiles increased, so did the rate of fatal accidents, particularly during the 1920s. This rise in death rates posed a public threat, sparking media debates and prompting authorities to take action. The solution was to keep streets clear of wandering pedestrians. Pedestrian traffic lights were among the multiple elements that helped achieve this goal.

During this period, the term "jaywalkers" emerged as a public shaming label for those who disregarded traffic light rules. The term "jaywalker" originated from the derogatory term "jay," which referred to a person from a rural area, implying that such a person was naive or dumb. In this context, a "jaywalker" was "someone who did not know how to walk in the city behave in a city environment, particularly with regard to traffic rules" (Norton 2014, 25). Here, the realm of the everyday is subjected to a new form of coexistence and behaviour with new technological artefacts in the public space. New artefacts can have serious implications for what we are and how we think of ourselves, both individually and collectively. However, more often than not, it is not made clear what the process of bringing a new technology into public use entails.

Noortje Marres's study of intelligent vehicle testing in the UK sheds light on how mundane trials in public spaces can be a double-edged sword. She views the street as a public trial space for introducing innovation to society, a process she describes as a "double-edged operation" (2020, 113). On one hand, it involves engaging with citizens and inviting them to interact with the new artefact, allowing designers to gather feedback and to learn from the public's use, opinions, needs, desires, and even rejection of the artefact. On the other hand, the operation involves the "normalizing effect" that the street trial has on the public's use of these new technologies. Consequently, the street trial is not just about learning from the invisibilities discovered through the artefact's engagement with the public, but also about making the artefact visible on the street to become part of everyday life.

In a similar vein, Sheila Jasanoff's (2016) *Ethics of Invention* illustrates the power of technoscientific inventions as a double-edged sword, one that makes our lives easier and more convenient while radically (re) shaping society, occasionally in an undemocratic fashion. In Jasanoff's account, traffic lights are "a reminder that technologies incorporate both expert and political judgments that are inaccessible to everyday users" (2016, 11). Ethical inquiry, from her purview, is a way to get behind the choices made during the technology's design – politically, scientifically, and materially. Continuing with the otherwise mundane traffic light as her example, she elaborates that

> a fatal accident at a regulated intersection raises issues of negligence and liability different from those raised by an accident at a crossing without lights. Running a red light is evidence of wrongdoing in and of itself because we have chosen to attribute legal force to the color red. (2016, 12)

As Jasanoff and Marres remind us, the introduction of any infrastructure of enforcement and control also becomes a device of law and regulation, which only leads to broader effects for everyday life. As soon as traffic lights are put in place, people are obliged to obey, regardless of how unjust the situation may be or when it makes no sense at all to stop. This marks a new demarcation between good citizen and bad, regardless of the conditions in which good and bad are performed.

In traffic scheduling, the dominant approach is to model a common traffic flow using software. Frey et al. argue that "human beings are far more adaptable to changing conditions than it is represented in the software" (2011, 73). The authors show that the current infrastructure of traffic management is not only inefficient but also unjust. It is primarily concerned with automobility and almost entirely misses the need to consider the elasticity of traffic to adjust to changing situations. In this line of thinking, Leth et al. (2014) advocate for jaywalking to be decriminalized for pedestrians and cyclists. Their proposal may offer a truly innovative approach for adaptive traffic modelling that differs from the contemporary smart technology narrative. To achieve this goal, the authors argue that it is public perception that must change.

Thus, the "smartification" of a mundane object like pedestrian traffic lights, while seemingly promising, often overlooks the complexities of human behaviour and societal structures. As Marres (2020), Sadowski (2020), and Jasanoff (2016) illustrate, the introduction of smart technologies, such as intelligent traffic lights, can have unintended consequences, often reinforcing existing power structures and creating new forms of control. This is not to dismiss the potential benefits of these technologies, but rather to call for a more nuanced and democratic approach to their design and implementation.

The case of intelligent traffic lights in Vienna, for instance, highlights the need for a more inclusive design process that considers the diverse needs and experiences of all city dwellers. It also underscores the importance of questioning the narratives and assumptions that underpin the "smartification" process. If, as Frey et al. (2011) argued above, human beings are far more adaptable to changing conditions than is often represented in software, we need to rethink our reliance on techno-solutionism and consider more flexible, human-centred approaches. In this regard, the proposal by Leth et al. (2014) to decriminalize jaywalking offers a compelling alternative to the dominant smart technology narrative. By challenging public perceptions and advocating for more adaptive traffic modelling, they highlight the potential for a more democratic and inclusive form of "smartification."

Importantly, the "smartification" of everything should not be an end in itself but rather a means to enhance the quality of life for all citizens. This requires a continuous process of critique, reflection, and engagement with diverse stakeholders. Moreover, "smartification" can serve as a moment of renegotiation about the forms of values and desired social relations rather than merely offering tech solutions to old problems. In essence, the "smart" solutions we develop should be truly smart, not just in their technological sophistication, but also in their ability to address the complex and often "dumb" problems that characterize our everyday lives, while fostering more equitable and inclusive social relations.

Conclusion: Reimagining Smartification

Traffic lights, ubiquitous devices that regulate and organize everyday life, often go unnoticed in their political implications. While they serve the seemingly mundane purpose of directing traffic, they implicitly prioritize automobiles over pedestrians. The smartification of pedestrian traffic lights, however, offers an opportunity to re-examine the broader sociotechnical assemblage of streets and mobility.

This chapter has explored how the new smart traffic light system in Vienna failed to consider the diverse needs and uses of the streets, particularly for visually impaired individuals who rely on the pushbutton box's speaker for navigation. The justification for introducing this new system in the city was based on a narrow framing of the issue at hand, focusing solely on replacing the pushbutton box without addressing the underlying problems of everyday mobility governance. The smartification of pedestrian traffic lights should not merely offer a technologically advanced solution to an old problem. Instead, it should provide an opportunity to rethink the status quo of pedestrian crossings and prioritize pedestrians in the design of traffic management systems. The initiative should be based on a comprehensive understanding of the issue, taking into account the conflicting priorities of different modes of mobility. The introduction of smart traffic lights opens a window to revisit the car-dominated mobility agenda of the twentieth century and approach it from a different perspective. The smart pedestrian traffic light needs to prioritize pedestrians and consider how technology can help achieve this goal rather than simply automating a problem that could be resolved by making the pushbutton more visible.

It is crucial to recognize that technological interventions are often driven by the imaginary of techno-solutionism and come with both intended and unintended consequences that increase control over social order. Therefore, it is time to reflect on why visually impaired people were neglected in the first place and how a more in-depth engagement with the multiplicity of use can be inscribed into the design protocol. This requires a sociotechnical understanding of the situation before implementing any urban initiative. Approaching the smart pedestrian traffic light in this vein, the inquiry into jaywalking shifts from being an individual misbehaviour to a sociotechnical interaction with the temporal ordering of mobility in relation to the mundane situations of streets and traffic lights as devices of control. In this context, the smartification of traffic lights should serve as a moment of renegotiation about the forms of values and desired social relations rather than merely offering tech solutions to old problems.

The evolving interpretations attached to ubiquitous pieces of urban infrastructure, such as traffic lights, underscore the complex ways in which their meanings can shift over time. These shifts are significant, particularly when considering the disparity between the intended and acquired meanings of such artefacts. The intended purpose of intelligent traffic lights in enhancing traffic efficiency and safety may diverge markedly from the meanings these systems acquire through their interactions with various social groups and urban dynamics. This disparity is notably pronounced among marginalized groups, such as the visually impaired, who may experience these infrastructures in fundamentally different ways than those envisioned by their designers.

For the visually impaired, intelligent traffic lights may not fully address navigational challenges or may inadvertently introduce new barriers, thus altering the meaning of these devices from facilitative to obstructive. This divergence in meaning highlights the necessity for a more nuanced understanding of how urban infrastructure intersects with the lived experiences of disabled populations. Exploring this dynamic provides an opportunity to connect the discourse on intelligent urban systems with broader public policy literature focused on disability and belonging. Such an integration not only enriches the analysis of smart urbanism but also contributes to ongoing discussions about inclusivity, accessibility, and the rights of disabled persons within rapidly modernizing urban spaces. This approach aligns with scholarly efforts to ensure that urban policy development is attuned to the diverse needs and experiences of all city dwellers, fostering a more inclusive urban environment.

In conclusion, the smartification of pedestrian traffic lights should be more than just a technologically intelligent solution. It should be an opportunity to reimagine the streets and mobility, prioritize pedestrians, and address the complex problems of everyday life in a more equitable and inclusive manner.

References

Ertler, Christian, Horst Possegger, Michael Opitz, and Horst Bischof. 2018. "An Intent-Based Automated Traffic Light for Pedestrians." 2018 15th IEEE International Conference on Advanced Video and Signal Based Surveillance (AVSS), Auckland, New Zealand, November 27–30. https://doi.org/10.1109/AVSS.2018.8639112.

Frey, Harald, Ulrich Leth, Anna Mayerthaler, and Tadej Brezina. 2011. "Predicted Congestions Never Occur. On the Gap Between Transport Modeling and Human Behaviour." *Transport Problems* 6 (1): 73–86. https://doi.org/10.13140/2.1.1928.5769.

Jasanoff, Sheila. 2016. *The Ethics of Invention: Technology and the Human Future*. W.W. Norton.

Joerges, Bernward. 1999. "Do Politics Have Artefacts?" *Social Studies of Science* 29 (3): 411–31. https://doi.org/10.1177/030631299029003004.

Leth, Ulrich, Harald Frey, and Tadej Brezina. 2014. "Innovative Approaches of Promoting Non-Motorized Transport in Cities." 3rd International Conference on Road and Rail Infrastructure – CETRA 2014, Spilt Dalmatia, Croatia, April 28–30.

Marres, Noortje. 2020. "Co-Existence or Displacement: Do Street Trials of Intelligent Vehicles Test Society?" *The British Journal of Sociology* 71 (3): 537–55. https://doi.org/10.1111/1468-4446.12730.

Mattern, Shannon. 2021. *A City Is Not a Computer: Other Urban Intelligences.* Vol. 2 of *Places Books.* Princeton University Press. https://doi.org/10.1515/9780691226750.

McShane, Clay. 1999. "The Origins and Globalization of Traffic Control Signals." *Journal of Urban History* 25 (3): 379–404. https://doi.org/10.1177/009614429902500304.

Norton, Peter. 2011. *Fighting Traffic: The Dawn of the Motor Age in the American City.* MIT Press. https://doi.org/10.7551/mitpress/9780262141000.001.0001.

— 2014. "Of Love Affairs and Other Stories." In *Incomplete Streets: Processes, Practices, and Possibilities,* edited by Stephen Zavestoski and Julian Agyeman. Routledge. https://doi.org/10.4324/9781315856537.

Sadowski, Jathan. 2020. *Too Smart: How Digital Capitalism Is Extracting Data, Controlling Our Lives, and Taking Over the World.* MIT Press. https://doi.org/10.7551/mitpress/12240.001.0001.

Sadowski, Jathan, and Roy Bendor. 2019. "Selling Smartness: Corporate Narratives and the Smart City as a Sociotechnical Imaginary." *Science, Technology, & Human Values* 44 (3): 540–63. https://doi.org/10.1177/0162243918806061.

Seghrouchni, Amal El Fallah, Fuyuki Ishikawa, Laurent Hérault, and Hideyuki Tokuda, eds. 2016. *Enablers for Smart Cities.* Wiley & Sons. https://doi.org/10.1002/9781119329954.

Sepehr, Pouya. 2024. "Mundane Urban Governance and AI Oversight: The Case of Vienna's Intelligent Pedestrian Traffic Lights." *Journal of Urban Technology* 40 (1): 1-18. https://doi.org/10.1080/10630732.2024.2302280.

Sepehr, Pouya, and Ulrike Felt. 2023. "Urban Imaginaries as Tacit Governing Devices: The Case of Smart City Vienna." *Science, Technology, & Human Values,* 50 (2): 364–86. https://doi.org/10.1177/01622439231178597.

Tironi, Martín, and Matías Valderrama. 2018. "Unpacking a Citizen Self-Tracking Device: Smartness and Idiocy in the Accumulation of Cycling Mobility Data." *Environment and Planning D: Society and Space* 36 (2): 294–312. https://doi.org/10.1177/0263775817744781.

Winner, Langdon. 1980. "Do Artifacts Have Politics?" *Daedalus* 109 (1): 121–36. https://www.jstor.org/stable/20024652.

Woolgar, Steve, and Geoff Cooper. 1999. "Do Artefacts Have Ambivalence: Moses' Bridges, Winner's Bridges and Other Urban Legends in S&TS." *Social Studies of Science* 29 (3): 433–49. https://doi.org/10.1177/030631299029003005.

Zeno-Zencovich, Vincenzo. 2016. "Lessons from a Traffic Light: A Juridical Scherzo." *European Journal of Comparative Law and Governance* 3 (1): 3–23. https://doi.org/10.1163/22134514-00301001.

20 Outsmarting Urbanism: Could Leveraging the "Right to Be Rural" Produce Alternative Futures?

S. ASHLEIGH WEEDEN

How Do You Solve a Problem like Rurality?

Approaching rural people and places – whether through research, policy, or practice – requires grappling with the complex, frequently opaque, and sometimes frustratingly elusive (if not actually illusive) process of defining where, what, and who is "rural." Rural spaces are "almost never conceived as simply space; they are almost always assigned and inscribed with meanings created by the human experience connected to those places" (Weeden 2022a, 40). These meanings, Woods (2010) argued, are often contested and conflicting; rural sites are conceived as idyllic, unruly, the location of resources, empty, undeveloped, backward, sites of opportunity, retreat, economic engines, and everything and anything in between – to varying degrees of accuracy or benefit to rural people and places themselves. What is valuable or meaningful, or what is at the centre and what is at the periphery, depends on who you are and where you're standing, making a definition of rurality highly dependent on why, exactly, a definition is required in the first place.

In particular, smartification agendas, and many of their associated activities, tend to assume one of two positions when it comes to rurality: rural people and places are either framed as behind, backward, or in need of "levelling up," and smartification is offered as the pathway to bringing rural places into line with their urban counterparts; on the other side of this ideological coin, rural areas are treated as resources and sites where otherwise dormant value should and must be extracted (typically to the benefit of urban people and places) and that smartification is the pathway for maximizing productivity. In either orientation, rurality is a problem to be solved.

Such problematics as a primary point of engagement have long been a source of frustration for rural people and practitioners. While there is a large and growing body of literature on the challenges and complexities of rural policy, the implications of assuming that rurality is a perpetual problem appears to have only just started gaining traction as an issue requiring attention. Researchers and practitioners have started noting that such treatment creates reinforcing feedback loops (see Markey et al. 2015; Sherry and Shortall 2019). To this end, Sherry and Shortall argue that rural places are often approached by "methodological fallacies" that produce evidence that reinforces the "assumption of rural disadvantage, the nature of which is never articulated" (2019, 336) and results in interventions "designed to treat disparity rather than accommodate diversity" (336). These arguments appear to echo and underscore an emerging critique of the "smartification of everything" as a disciplining discourse that equates the "virtue" of being "smart" with technology and positions all social complexity as a problem to be solved by that technology. When layered together, the arguments baked into assumptions about rural disadvantage and the promises of smartification seem to indicate that, in many spheres, successful rural development looks like urbanization (see Bollman 2019; Weeden 2020). If rural people and places are problems – seen as failed cities, cities in waiting, or resource banks for urban demands – then the overarching message from academic and policy research combined appears to be that smartification and urbanization are the solution to the problem of rurality.

The goal of this chapter, then, is to flip the script in a purposefully exploratory and transgressive manner by considering two key questions: (1) what happens if we problematize smartification and centre rurality? (2) Is there an opportunity to develop a "toolkit" or conceptual framework for critiquing rural smartification? The following sections briefly outline the key concepts and ongoing contestations inherent to rurality, smartification, and rural-urban dynamics before integrating them into a proposed framework to test against future empirical investigations into "the smartification of everything" in rural contexts.

The Right to Be Rural in the Shadow of Geographical Narcissism

Whether one takes a geospatial, demographic, political, economic, sociocultural, phenomenological, or temporal approach to defining rurality depends on who you are, where you come from, and what kinds of questions you are considering – which results in multiple, contested, and often conflicting ideas about rurality (see du Plessis et al. 2001; Woods 2010). Rural sites are conceived as idyllic, unruly, the location of resources, empty, undeveloped, backward, sites of opportunity, retreat, economic engines, and everything and anything in between – to varying degrees of accuracy or benefit to rural people and places themselves. Commonly, however, rurality is framed as a problem to overcome, rather than something to understand or something unique and valuable in and unto itself, with rural people seen as authorities on their own experiences, and entitled to their own unique realities and pathways (see Bollman 2019; Markey et al. 2015; Weeden 2020, 2022a, 2022b).

In the last five years, the conflict between rurality-as-concept and rurality-as-material-reality has generated work on what psychologist Malin Fors (2018) refers to as "geographic narcissism." Practicing in the world's northernmost town, Hammerfest, Norway, Fors argued that urban norms of power, space, place, and time are reinforced through professional practice, research, governance, and policy in ways that could be considered colonialization. This argument echoes critical interrogations of globalization as an ideology that "masks a hegemony of developed nations, powerful over those still regarded as developing" (Balfour et al. 2008, 96); "geographic narcissism" can and does operate at multiple scales, from the nation-state to the neighbourhood, whenever one place's way of living is positioned as the superior blueprint to all others. As a concept, "geographic narcissism" neatly captures the frustrations that run through much rural research, practice, and advocacy. When urbanity is assumed as not just the default, but the preferred and even superior mode of socioeconomic organization and operation, rural actors often must justify their experiences and relationships as valid and valuable under the forced and false binary of rural-urban comparison. The concept of "geographic narcissism" also opens the door to critically examining where and how knowledge about rural places is created and contextualized. This chapter is grounded largely in the Canadian context as this is the context for my own experience and knowledge, and it also presents an interesting divergence from the strong influence of American and European scholarship in socio-technical studies. Here, too, critical consideration of embedded relational geographies is required to surface the "geographic narcissism" within research, rural and otherwise, wherever it occurs (see Rignall and Atia 2017).

In response to the perceived implications of "geographic narcissism," the provocative notion of the "right to be rural" has emerged in the last decade, notably through the work of Barraclough (2013) and Foster and Jarman (2021). It is a response to and an extension of Lefebvre's (1996) concept of the "right to the city" that uses a rights-based lens to place rurality on equal footing with urbanity; it is neither better nor worse, but different. This differentness is shaped by spatial-demographic considerations, such as low population density and high distance to density, but it also includes social, political, cultural, economic, and environmental factors. These various qualities also produce significant diversity among and between rural people and places themselves. Using a "rights-based" approach highlights the ways in which rural people or places may require different supports or interventions than their urban counterparts to achieve equitable outcomes. Bollman (2019) argued that this recognition, framed as "differentiated universalism," is the key to successfully realizing the "right to be rural." As such, rurality as both a concept and a lived experience may offer an interesting and yet underexplored lens for confronting the hegemonic assumptions of standardization and urbanization that currently dominate both research and policy practice. In this way, and as similarly argued through different angles and applications in this volume (such as Moreschi and Pereira's "Positive History of NO," or Macktoom and Fatima's exploration of just how silly and poorly constructed western "smartness" seems when applied out of context), exercising a context-specific "right of refusal" to "smartness" may be one way that places that exist outside the imagination of the western city may be able to reclaim and retain the ability to say "yes" to their own agenda.

Urbanization and Smartification as Disciplinary Discourses

When "smart city" concepts are extended to policy agendas and industrial or sectoral development strategies, they act as an extension of what Vanolo (2014) conceptualized as a disciplining "smartmentality." This concept describes how "smartness" is suggested as good, natural, univocal, and inevitable, while acting as a tool for the political legitimization of capitalist interests in technological applications. Further, Vanolo argued that the terms used by "smart city" proponents represent "evocative slogan[s] lacking a well-defined conceptual core and, in this sense, proponents of the smart city are allowed to use the term in ways that support their own agendas" (2014, 884), making it critical to examine who is setting the agenda. The production of "smartmentalities" via smartification represents a disciplinary discourse and practice because the purpose of such projects is driven, framed, and understood solely through the rationalities and goals of the most powerful stakeholders involved – typically large, multinational technology companies (like Cisco Systems, Google's parent company Alphabet, Microsoft, IBM, and others like these; see Sadowski and Bendor 2018).

Although not fully explored, it appears that "geographic narcissism" is baked into Vanolo's (2014) concept of "smartmentalities," which is myopically focused on urban applications of smartification; the combination of urbanity and technology offers a remedy to the unruliness of everything else. However, neither urbanization nor technology are neutral nor inevitable, but rather directly shaped by and a reflection of the choices and values of specific individuals in decision-making positions or positions of influence and power. As a result, "smartmentalities" and "geographic narcissism" combine to produce smartification agendas that serve to discipline rural regions. Smartification agendas are designed to enforce and achieve urban assumptions about productivity, agglomeration benefits, and spatial efficiency through the application of standardized technology. In the Canadian context, Spicer et al. examined the "frontier" of smartification in the Annapolis Valley and in Iqaluit, concluding that the "rapid adoption of smart city technology is leading to a tiered system of 'haves' and 'have nots'" (2019, 16), where the conflation of smartness with development and urbanized standards as the arbiters for quality of life are noted as a challenging tension for rural communities with aspirations of smartification. Additionally, in their evaluation of Canada's Smart Cities Challenge, Goodman et al. (2020) note that the process set out to drive the adoption of smartification by Canadian municipalities appears to have re-embedded top-down policy decisions that struggle to reconcile the complexities of place-based public engagement with smartification. In both articles, smartification is still positioned as a worthy, even inevitable, pursuit. The promised outcome of smartification is that rural disadvantage can be overcome by adopting technologically oriented interventions and finding workarounds to compete with urban counterparts. Such promises indicate that, with the right disciplining strategy, rural regions can access or achieve the same, presumably positive, outcomes as cities. Such dynamics highlight Fors's (2018) argument that "geographic narcissism" may, in fact, be part of the ongoing project of colonial discipline enacted by the dominant economic class on areas that do not conform to their ideals. As such, viewing "smartmentalities" through a rural lens highlights the influence of both governments and large, multinational companies in leveraging technology to extend hegemonic urbanism. These powerful interests use smartification agendas to assimilate people and places into obedient subjects – but what happens to places that cannot (or will not) conform?

Rural regions may struggle to conform to the ideals of urban-based "smartmentalities" because they require certain variables to be in place in order to achieve their disciplinary goals. The broad range of rural environments (from pastoral to Arctic and everything in between) and their low density and high distance to other population centres make installing standardized, urban-developed, physical, technological infrastructure expensive and challenging to maintain, which, in turn, shrinks the return on investment for corporations. Rural regions are also often inconvenient for large urban firms to send representatives to; they may be far from transportation centres, lack accommodations, or be inaccessible during different times of year. Finally, rural institutions are often capacity-stretched, with limited resources (human and financial), and simply may not have the interest or ability to support the demands of large-scale technological intervention (see Flora et al. 2016; Weeden and Kelly 2021). Where civic smartification agendas often depend on the cooperation and support of civic institutions to prop up the goals of individual technology firms, they may find no purchase in rural regions that do not have authority, resources, or capacity to invite, evaluate, or manage the kind of relationships that appear common in "smart city" projects (see McCord and Becker 2019; Spicer et al. 2019). Further, because there is often no clearly articulated and commonly desired goal of a civically oriented smartification agenda (see Cole et al. 2022) – other than to advance the interests of capitalist firms – it is difficult to construct a logical process for using such blunt instruments to meet locally specific needs.

When "geographic narcissism" goes unchecked and is bolstered by "smartmentalities," urbanized norms and power structures are assumed to exist everywhere. When those layered assumptions encounter the reality of rural diversity, the order and control promised by smartification runs directly into the unruliness of rurality, and in response, rurality is framed as the reason for failed or challenged smartification. Rarely are rural people and places encouraged to challenge smartification and carefully consider what, why, and how they wish to implement technological interventions to civic or socioeconomic challenges. It is this last point that seems most promising and challenging for advancing a critical research agenda regarding smartification in rural contexts, and it is my hope that this chapter provides the conceptual encouragement required to begin exactly such an empirical undertaking.

Outsmarting Urbanism Through the Right to Be Rural: Producing a Critical Framework for Critiquing Smartification in Rural Contexts

As discussed above, rurality is unruly and resists containment, both in theory and in lived experience. Rurality is spectacularly diverse both within given nation-state jurisdictions and around the globe, producing the well-worn colloquialism: "if you know one rural community ... then you know one rural community" (Markey et al. 2015, 2; see also Breen et al. 2021). These complex and contested undercurrents make rural research a diverse, multi- and trans-disciplinary effort. They are also critical to understanding why the diversity and complexity of rural people and places can, and could, be approached as sites of conceptual and material resistance to the overall flattening effect of neoliberalism and late-stage capitalism. As argued earlier, smartification agendas represent exactly this phenomenon, as a means to discipline social relations through neoliberal, capitalist applications of technology.

When it comes to many of the activities and strategies connected to smartification agendas, such as digital infrastructure, digital skills, innovation, and technological assets, rural people and places are still treated as deficient or disadvantaged (see Spicer et al. 2019). This is largely because rural regions have typically been on the lagging end of investments in the technological infrastructure required to support smartification, producing a "digital divide" that is both material (e.g., a lack of broadband infrastructure) and intangible (e.g., skill or interest) (Weeden and Kelly 2021). If the "right to be rural" movement has gained any traction in this space, it is typically in service of arguing that rural people require and deserve the same quality of digital infrastructure and services as their urban counterparts; the concept is rarely, if ever, used to support efforts by communities to slow down any technological investments. While these need not be zero-sum positions, it is troubling to sense that the invocation of the "right to be rural" in terms of technology may get stuck at re-embedding dynamics where rural regions continue to see themselves through urban lenses, rather than advancing a movement where rural regions develop the capacity to identify, pursue, and achieve relationships with technology that serve self-determined, self-governed rural interests.

This interesting tension between rurality and the "right to be rural" as concepts versus the way rurality is experienced and leveraged in reality presents an important entry point for inviting researchers and communities to reconsider approaches to policy and practice. Instead of viewing the complexities of rurality as a problem to overcome by accepting whatever technology is offered by private corporations, rural diversity offers an opportunity for being specific and purposeful about the adoption and integration of technology and smartification. Centring rural realities, rather than attempting to discipline them through urbanized "smartmentalities," could reveal new ways for communities to leverage critical engagement with smartification. This could, in turn, produce more effective place-based policy and programming that follows, rather than directs, locally driven strategies for working with technology. Doing so requires a new framework for surfacing the power dynamics and political economy of smartification in rural contexts.

Below, I propose a high-level, preliminary sample of such a framework that leverages the concepts explored in this chapter, offering critical questions for engaging in constructive critique of smartification in future empirical research on rural contexts:

Table 20.1. Framework for Critiquing Smartification in Rural Contexts

Disciplining Themes/Questions to Surface Critique		Leveraging the "Right to Be Rural" as Counter-Critique/Centring Discourse to Examine Smartification
Smartmentalities	Geographic Narcissism	Right to Be Rural
Promises of technology: social, economic, future-opportunities	Urban-by-default, assumed rural disadvantage (implied/explicit)	How will this "smart" activity or agenda support rural needs/goals? Does this technology facilitate the access or activation of other rights?
Technology as a solution to rurality	Location/orientation of proponent (rurally oriented vs. urban-based)	Does the technology or agenda "work" in a rural context? What would be needed to adapt it to rural contexts? Is that possible or desirable? Do those adaptations change the result of adopting the technology? What are the implications of refusal or revision of a proposed technology?
Private investments promised/promoted as substitute for public oversight and infrastructure	Standardized vs. place-specific	What challenges might the technology pose to accessing or exercising some rights? Who has the right to agree or disagree? How is consent managed? How will the technology be governed?
Technology as inevitable, "good," moral imperative	Goals of development	What trade-offs or costs are involved in accepting or adopting this technology? What happens if there are negative impacts – how will they be managed and by whom? Who will bear the costs?
Visible (explicit) and invisible (implicit) norms vs. lived realities/ experiences	Visible (explicit) and invisible (implicit) norms vs. lived realities/ experiences	Is rurality considered on its own or in comparison? In what ways? How is rurality visible? How will success be evaluated?

Summary: Imagining Alternative Futures

Imagining alternatives to the "smartification of everything" requires transgressing deeply entrenched assumptions about technology, urbanity, rurality, and power. Considering things from the "right to be rural" framework requires centring rurality on its own terms, which produces observations that indicate rurality may be a site of resistance to "smartmentalities." The degree to which such resistance or engagement is happening, and how, offers fertile ground for empirical investigation, as this chapter represents the initial conceptual exploration of questions that do not yet appear to have been fully considered from a rurally centred or "right to be rural" orientation.

The work of considering whether and how such resistance is possible or occurring is best facilitated by going back to the first principles of good public policy and governance: centring those most impacted by a given decision or policy. The framework I have proposed for critically engaging with smartification in rural regions invites anyone undertaking this work to consider who is driving the agenda, what is said and what is left unsaid, how it will be evaluated, and by whom. Such questions quickly unravel "smartmentalities," which use the convenience, proximity, and density of urbanity (which serves to create distance between residents and decision-makers) to avoid clearly articulating the goal of smartification – which, when exposed, reveals itself to be the advancement of the economic interests of private capital (see Cole et al. 2022; McCord and Becker 2019; Sadowski and Bendor 2018).

In order to work in and for rural regions, smartification has to adapt to rural realities through diverse and specific applications – which runs directly contrary to the disciplinary nature and "geographic narcissism" of "smartmentalities." It is, however, up to rural people themselves to see this conflict as an opportunity to exercise their agency and to determine their own agendas for whether, why, and how they wish to accept or adopt technological interventions in civic life.

References

Balfour, Robert J., Claudia Mitchell, and Relebohile Moletsane. 2008. "Troubling Contexts: Toward a Generative Theory of Rurality as Education Research." *Journal of Rural and Community Development* 3 (3): 95–107. https://journals.brandonu.ca/jrcd/article/view/139/49.

Barraclough, Laura. 2013. "Is There Also a Right to the Countryside?" *Antipode* 45 (5): 1047–9. https://doi.org/10.1111/anti.12040.

Bollman, Ray. 2019. "Right to Be Rural: Rurality as a Case for Differentiated Universalism." Paper presented at the 2019 Annual Canadian Rural Revitalization Conference, St. John's, Newfoundland, Canada, October 1–5. Canadian Rural Revitalization Foundation. www.dropbox.com/s/pc52ovjvto67tz2/CRRF2019Bollman.pptx?dl=0.

Breen, Sarah-Patrice, Lars K. Hallstrom, S. Ashleigh Weeden, Sean Markey, and Bill Reimer. 2021. *Building the Canada We Want in 2050: Submission to Infrastructure Canada*. Canadian Rural Revitalization Foundation. http://crrf.ca/building-the-canada-we-want-in-2050/.

Cole, Alistair, Calvin Lai Ming Tsun, Dionysios Stivas, and Emilie Tran. 2022. "Chapter 7: The 'Smart City' between Urban Narrative and Empty Signifier: The Case of Hong Kong." In *Constructing Narratives for City Governance*, edited by Alistair Cole, Aisling Healy, and Christelle Morel Journel. Edward Elgar. https://doi.org/10.4337/9781800374454.00013.

du Plessis, Valerie, Roland Beshiri, Ray D. Bollman, and Heather Clemenson. 2001. "Definitions of Rural." *Statistics Canada Rural and Small Town Analysis Bulletin* 3, no. 3 (November). Catalogue no. 21-006-XIE. https://www150.statcan.gc.ca/n1/pub/21-006-x/21-006-x2001003-eng.pdf.

Flora, Cornelia B., Jan L. Flora, and Stephen P. Gasteyer. 2016. *Rural Communities: Legacy + Change*. 5th ed. Routledge. https://doi.org/10.4324/9780429494697.

Fors, Malin. 2018. "Geographical Narcissism in Psychotherapy: Countermapping Urban Assumptions About Power, Space, and Time." *Psychoanalytic Psychology* 35 (4): 446–53. https://doi.org/10.1037/pap0000179.

Foster, Karen, and Jennifer Jarman, eds. 2021. *The Right to Be Rural*. University of Alberta Press. https://doi.org/10.1515/9781772125955.

Goodman, Nicole, Austin Zwick, Zachary Spicer, and Nina Carlsen. 2020. "Public Engagement in Smart City Development: Lessons from Communities in Canada's Smart City Challenge." *The Canadian Geographer* 64, no. 3 (Fall): 416–32. https://doi.org/10.1111/cag.12607.

Lefebvre, Henry. 1996. *Writings on Cities*. Translated and edited by E. Kofman and E. Lebas. Blackwell.

Markey, S., S. Breen, R. Gibson, A. Lauzon, R. Mealy, and L. Ryser, eds. 2015. *The State of Rural Canada: 2015*. Canadian Rural Revitalization Foundation. https://sorc.crrf.ca/wp-content/uploads/2015/09/SORCExSum.pdf.

McCord, Curtis, and Christoph Becker. 2019. "Sidewalk and Toronto: Critical Systems Heuristics and the Smart City." In *Proceedings of the 6th International Conference on ICT for Sustainability*, edited by Annika Wolff. http://ceur-ws.org/Vol-2382/.

Rignall, Karen, and Mona Atia. 2017. "The Global Rural: Relational Geographies of Poverty and Uneven Development." *Geography Compass* 11 (7): e12322. https://doi.org/10.1111/gec3.12322.

Sadowski, Jathan, and Roy Bendor. 2018. "Selling Smartness: Corporate Narratives and the Smart City as a Sociotechnical Imaginary." *Science, Technology, & Human Values* 44 (3): 540–63. https://doi.org/10.1177/0162243918806061.

Sherry, Erin, and Sally Shortall. 2019. "Methodological Fallacies and Perceptions of Rural Disparity: How Rural Proofing Addresses Real Versus Abstract Needs." *Journal of Rural Studies* 68 (May): 336–43. https://doi.org/10.1016/j.jrurstud.2018.12.005.

Spicer, Zachary, Nicole Goodman, and Nathan Olmstead. 2019. "The Frontier of Digital Opportunity: Smart City Implementation in Small, Rural and Remote Communities in Canada." *Urban Studies* 58 (3): 535–58. https://doi.org/10.1177/0042098019863666.

Vanolo, Alberto. 2014. "Smartmentality: The Smart City as Disciplinary Strategy." *Urban Studies* 51 (5): 883–98. https://doi.org/10.1177/0042098013494427.

Weeden, Sara Ashleigh. 2020. "Will Post-COVID Policies Realize the Full Potential of Rural Canada?" *Policy Options*, July 8. https://policyoptions.irpp.org/magazines/july-2020/will-post-covid-policies-realize-the-full-potential-of-rural-canada/.

– 2022a. "Place, Power, and Policy in the 'Nuclear North': A Critical Comparative Analysis of Policy Narratives about Rural Innovation Systems Anchored by the Nuclear Energy Sector in Scotland and Canada." PhD diss., University of Guelph. https://atrium.lib.uoguelph.ca/items/5f8d2bf1-8650-4751-bdd3-8a7d02180192.

– 2022b. "The Right to Multiple Futures in the Shadow of Canada's Smart City Movement." In *The Right to be Rural*, edited by Karen Foster and Jennifer Jarman. University of Alberta Press. https://doi.org/10.1515/9781772125955.

Weeden, Sara Ashleigh, and Wayne Kelly. 2021. "Canada's (Dis)connected Rural Broadband Policies: Dealing with the Digital Divide and Building 'Digital Capitals' to Address the Impacts of COVID-19 in Rural Canada." *The Journal of Rural and Community Development* 16 (4): 208–24. https://journals.brandonu.ca/jrcd/article/view/2057.

Woods, Michael. 2010. *Rural*. Routledge. https://doi.org/10.4324/9780203844304.

21 Rural Expertise and the Sewer

JEAN HARDY

In April of 2019, I sat down with an employee of the city of Houghton, Michigan, population 8,386 in the 2020 Census, to discuss ongoing economic development work in the city and the surrounding region. I was in my eighth month of living in the region as part of ongoing ethnographic research on rural development efforts in Michigan and how they were being transformed in a world obsessed by the high-tech economy and increased emphasis on innovation at all costs. The office where we met is located in the gorgeous red sandstone building pictured below, the Houghton City Center building. Built in 1905 out of locally extracted Jacobsville Sandstone, a type of sandstone only found in this area of the world, the building was originally used as a Masonic Temple but was taken over by the city in 1989 to preserve its use and maintenance in a time of economic decline.

Mining was the primary industry of Houghton and the surrounding region (i.e., the Keweenaw Peninsula) for over one hundred years, from approximately 1850 to the 1960s, with the region being home to one of the largest deposits of native copper in the world. The last copper mine closed in 1968. The city employee I interviewed was born around the time of the last mine closure, in a former mining town about twenty miles north of Houghton. Encouraged by his family to get a university degree and get out of the dying mining town he grew up in, he moved to Houghton, got an engineering degree at the local university, and worked as an engineer before beginning a career in local government.

A few years prior to our interview, the city of Houghton received a US$500,000 grant from the State of Michigan to improve local infrastructure data. The grant was meant to help low-resource communities develop asset management plans for their stormwater and wastewater systems. In other words, this program was meant to help communities like Houghton develop new digital systems using video equipment and

Figure 21.1. The Houghton City Center building, located in downtown Houghton, Michigan. Photo by Betsy Lehman, used with permission by the photographer.

geographic information systems (GIS) to survey the quality and health of sewer and wastewater infrastructure. Such digital system development would promote equitable access to new digital tools and improve infrastructure maintenance and management. Yet, by the time of our interview in April 2019, only two years after the project finished, the digital system and accompanying data created by its development had already fallen out of use.

So, what happened? Why did this rural municipality, and other communities I became familiar with through my fieldwork, take advantage of this opportunity from the state, but not use the new data-driven intervention? This state initiative had the signs of a successful government program: almost everyone that applied received the grant; there was no matching requirement for the grant, meaning that the communities on the receiving end didn't need to dedicate their limited municipal funds to ensure its receipt; and the grant was designed to promote equity in infrastructure development and maintenance.

What I demonstrate in this chapter is how, in this specific instance, the deployment of new forms of digital data and their accompanying systems butted up against the local and rural expertise of the sewer. In other words, the promised smartness of new data-driven systems wasn't enough to replace the labour and expertise that already existed in the community. The small scale of the system encouraged a familiarity with it, held by municipal employees, rather than in new forms of digital data. Recognizing this helps us understand the impact, or lack thereof, surrounding this rural tech intervention. This case argues, in part, for a realization that smartness is context dependent. The infrastructural intervention, despite its equitable strategy, did not succeed because of a generalized approach that expected smart systems to operate and provide benefits to municipalities divorced from their geographic and workplace context. As the editors of this volume articulate in the introduction, smartification is often a process of "rendering ... complex social relations and environments technical or 'smart'" (Gugganig et al., 6). In part what I argue here is that smartification can also introduce unneeded complexity into a situation that is already understood, rendering the new "smart" technology useless from the start. The remainder of this chapter unfolds as follows: first, I briefly introduce the field site and inspiration at the centre of this case; second, I explain what asset management of infrastructure means and how this materializes in one particular program in Michigan; third, I share examples from my fieldwork in two rural municipalities where I encountered this failure of adoption; finally, I reflect on the role of embodied expertise in rural work and how this expertise butts up against the smartness encoded in digital data.

Field Site and Background

What I present here is one small piece of a broader project that seeks to understand the role of contemporary digital technology and data in rural development discourses and practice. My research is interested in understanding those that advocate for rural opportunity and inclusion in digital futures. To understand this, I've been conducting ethnographic research in the Upper Peninsula of Michigan since 2014. In 2018 and 2019, I spent eighteen months living and researching in one particular region of the Upper Peninsula called the Keweenaw Peninsula. During my ethnographic research, I interviewed economic developers, municipal leaders, and local entrepreneurs. I embedded myself in the work of two economic development organizations that were serving some of the most remote and low-resource communities in the region. I attended municipal and county committee meetings, the events of local start-ups and business incubators, and spent time each week working out of a local co-working space recently created by a pair of local entrepreneurs and remote workers. I initially learned about the state's sewer and wastewater asset management program as I was interviewing about the types of digital technologies that economic development organizations and municipalities were using in their work. Early interviews with economic development officials resulted in discussions around the growing role of GIS in tracking infrastructure, and I incorporated this into my interview protocol with city officials as the research progressed. What is included in this chapter is drawn from various data collection opportunities, including interviews with current and former city managers of two municipalities, interviews with multiple economic development staff, and background research on the implementation of asset management policies in Michigan.

While most of the inspiration for this piece is drawn from my time living and working among the residents of the Upper Peninsula who were navigating the complex resource and technological landscape of their rural region, I also draw inspiration from important scholarship on technology and rural development. Research on the efforts to transform rural regions, bring them "up to date," make them "smart," and to create economic opportunity are not a new phenomenon. For example, rural sociologists John Allen and Don Dillman (1994) wrote about the impact (or lack thereof) of the "information age" on a rural town in Washington State in the 1980s. Historian Ronald Kline wrote extensively (e.g., Kline 2000) about the introduction of new technologies in rural America in the early twentieth century, and how rural residents actively shaped that adoption process through appropriation as well as rejection.

I also find inspiration in the critical computing scholars who investigate the relationships between rural people, traditions, and industries on one side, and the design and use of digital systems on the other. For example, Steup et al. (2019) write about the explosion of agricultural start-ups working to peddle their visions of a data-driven farm future and how this might impact power relations between farmers and other stakeholders in agriculture. Phoebe Sengers's extensive work on time and datafication of rural traditions in the North Atlantic (e.g., Sengers 2011; Sengers et al. 2021) documents how traditions clash with contemporary digital expectations. Drawing from all of this, I come to this work with a critical eye to how digital systems are created by and for rural people rather than with them, and how this is reflected in their design and use (Hardy 2019; Hardy and Lindtner 2017; Hardy et al. 2019).

Asset Management and the Digitization of Rural Infrastructure Data in Michigan

Through most of the twentieth century, municipal sewer and wastewater systems, especially in rural areas, relied on paper records to track the locations and other data associated with their infrastructure. As computer-based GIS advanced, so did their adoption in mapping municipal infrastructure. Starting in the 1980s, engineers and governments started to use the language of "asset management" to refer to the management of publicly owned infrastructure (Cagle 2003).

While asset management has multiple meanings, including a related use for the management of financial assets, asset management in this case deals with the management of infrastructural assets, such as roads, sewers, and power lines, that are public forms of infrastructure in which a government has made a long-term investment. As Cagle states, "asset management is embodied in knowing ...What you have; What condition it is in; What the financial burden will be to maintain it at a targeted condition" (2003, 1–2). Through the process of asset management, infrastructure is evaluated on those types of measures and that data is collected, mapped, preserved, and maintained in GIS. It is often done in a way that standardizes the data and allows for interoperability among regional, state, and federal bodies (Baird 2011). To summarize, asset management allows for both the mapping and tracking of publicly owned infrastructure and develops digital systems and data to serve those functions.

The State of Michigan has utilized asset management to track infrastructure it owns since 1997. Its sewer and wastewater asset management (SAW) program was started in 2013. Unlike the transportation asset management program, which focused on infrastructure that was owned by the state, this program focused on developing asset management programs across hundreds of municipal sewer and wastewater systems in Michigan. While the larger cities in Michigan had existing asset management programs for their water infrastructure by 2013, rural communities and low-resource small cities did not. Through the program, US$450 million was allocated, and cities, counties, townships, and publicly owned utility companies were encouraged to apply. The state distributed six rounds of support through 2020, and all of the entities in my field site that were eligible and applied received funding.

The overall goal of the SAW grants was to modernize the maintenance of public infrastructure through the creation and central interoperable storage of digital data about sewer and wastewater infrastructure. In one ideal case from the blog of ESRI (Mann 2018), the corporation who owns the GIS products that are most often used to visualize the data (i.e., ArcGIS), a township used its SAW funds to digitize its wastewater network so that it could streamline maintenance and help it respond to potential emergencies. ESRI's blog post framed this as a digital upgrade for the township, that the adoption of the asset management system "moved [the town's] outdated manual processes to modern automated workflows" (Mann 2018). The post continued to share that township staff now had a mobile app that the team uses to monitor its water system and provide contextual data when things do go wrong.

"It Gives More Data, but It Doesn't Necessarily Make Things Quicker"

The vision provided by ESRI was the purported goal of the asset management program: to use newly created digital data and visualization platforms to create efficiencies and modernize workflows. But this vision of a data-driven infrastructural future was not necessarily realized on the ground in rural Michigan.

By the time I was able to interview city leaders in my field sites in early 2019, all had completed their grants and submitted their final reports to the State of Michigan. The city of Houghton used the SAW grant to compile a digital inventory of all sewer system assets, craft an asset management plan, and hire a contractor to develop a GIS system for data tracking. Through this process, the city determined that nearly a third of its sewer and wastewater infrastructure was in poor condition. The funds from the SAW program were only allowed to be used to build an asset management system, not fix problems that it found. Houghton was required to outline how it was going to address these deficiencies over the next twenty years at a cost of US$8 million it didn't have. The city of Hancock undertook a similar assessment and found that 46 per cent of its main sewer lines were in poor condition. They estimated that the total costs to address the findings would be nearly US$10 million, money they did not have. In other words, both rural communities found significant issues with their sewer and wastewater systems, and both would need to borrow heavily to resolve the issues.

Interviews with city leaders revealed a much different story than the upbeat portrayal on ESRI's blog, in which the seamless integration of new mapping technologies into daily work was relatively simple. Instead, both city leaders emphasized that the implementation of a GIS stemmed more from the needs of the engineering firms they hired rather than the genuine necessities of the Department of Public Works (DPW) employees responsible for sewer maintenance within the two small towns. As one of them said,

> The value of GIS is oftentimes more for the engineers and not for the municipal DPW crew. My DPW crew is one main water person and one main sewer person and they know the system. They don't need to look it up.

In other words, the GIS was a way of codifying local knowledge for contracted engineers that was already largely known and operationalized in the work of the DPW employees. In the case of these neighbouring small rural cities with a combined population of only ~12,000 people, the infrastructure that existed was small enough, and their employees familiar enough, that the new asset management systems were seen as largely useless, an artefact of a grant program that wasn't incorporated into their work. As one of the city leaders told me, the laptop with the data and GIS system was sitting in a filing cabinet down the hall from his office.

Another city leader I spoke to explained to me how he perceived the potential of asset management technology and GIS "would be the cure-all for efficiency and quick access to hydrants and valves." He told me that, in the way the systems were portrayed by the state, "if you have a water leak, you could immediately get the guy called out and he could, based on his laptop, go right to the water break and shut it off." However, he continued, "it's not as practical as operationalized." Rather, he explained a complex process and said that it might be possible to identify the locations of the breaks quicker, but that the work of the DPW employees would not actually get completed any quicker. Instead, the familiarity the employees already had with their small sewer and wastewater system were what gave them the knowledge they needed to quickly identify where the issues would occur.

In his words: "It gives more data, but it doesn't necessarily make things quicker."

Butting Up Against Rural Expertise in a Time of Digital Infrastructure Data

As I and others have argued, information and communication technologies are designed for particular scales of intervention. In the case of privately owned technology companies, scale is required for the extraction of capital and creation of profit (Hardy 2019). The promises of technology itself as a scaling mechanism (Avle et al. 2020) that allows bureaucrats to see infrastructure at scale through digital systems is part of what makes systems such as asset management so attractive. But the promise of scaling breaks down in these cases, when small-scale, largely rural infrastructure is the site of intervention.

Despite the promised potential for transforming how municipalities maintain and track infrastructure, in the case presented in this chapter, the systems were not used as intended. Instead, the SAW data sat on laptops that were purchased as part of the grants, only to be used and updated by the engineering firms as part of their contract work. As S. Ashleigh Weeden argues in the chapter preceding this one, digital technologies are often leveraged as a way to solve problems of the rural without interrogating how rural communities are centred in the process and how rural needs/goals might actually be addressed. In the case of the implementation of Michigan's SAW program, the program attempted to account for potential concerns with respect to capacity building and equity: there were no requirements for external matching funds, most applicants received the funds, and the program targeted places (rural and urban) who couldn't otherwise afford these systems. Yet, as Weeden similarly argues, there was seemingly little to no consideration as to whether or not these types of systems were appropriate, would "work," or could functionally scale to such small infrastructural systems with any real use. One major outcome of the state's SAW program should have been a democratization of digital infrastructure data access and use. But what we see here is that it doesn't matter if communities have access to the data if they don't have the funds or needs to utilize it as well. Instead, they fall back on the localized expertise of their employees, their knowledge of those small-scale sewer systems, to continue the work.

By drawing attention to the context in which "smart" systems are deployed, we come to understand much more about the labour, and unwritten expectations of change in labour, that go into the adoption of new digital systems. We also learn about the universal assumptions baked into models of "smart" systems that seek to digitize the workings of municipal governments. In this case, the "smart" system was rendered by the state as the best new alternative, and funding was provided for equitable adoption, yet no attention was given to the context in which new digital systems would be deployed. But what might alternatives be? There are many unique opportunities for rural communities to use digital data and civic tech efforts in combination with community expertise to combat some of the biggest issues they face today. But, as I've demonstrated here, it's not enough to just fund communities with existing digital interventions. As prior research in development has argued, a development project is a complicated and multifaceted endeavour that requires many types of expertise (Brand and Karvonen 2007). Brand and Karvonen (2007) advocate for alternative forms of expertise, such as the meta-expert, whose job it is in the development process to broker multiple types of expertise, acting as a bridge builder between different parties of experts. Moving forward with rural smart systems, if we are to at all, means designing and implementing systems with multiple types, scales, and relationships to data and expertise in mind.

References

Allen, John C., and Don A. Dillman. 1994. *Against All Odds: Rural Community in the Information Age.* Westview Press.

Avle, Seyram, Cindy Lin, Jean Hardy, and Silvia Lindtner. 2020. "Scaling Techno-Optimistic Visions." *Engaging Science, Technology, and Society* 6:237–54. https://doi.org/10.17351/ests2020.283.

Baird, Gregory M. 2011. "Defining Public Asset Management for Municipal Water Utilities." *Journal – AWWA* 103 (5): 30–8. https://doi.org/10.1002/j.1551-8833.2011.tb11449.x.

Brand, Ralf, and Andrew Karvonen. 2007. "The Ecosystem of Expertise: Complementary Knowledges for Sustainable Development." *Sustainability: Science, Practice and Policy* 3 (1): 21–31. https://doi.org/10.1080/15487733.2007.11907989.

Cagle, Ron F. 2003. "Infrastructure Asset Management: An Emerging Direction." *AACE International Transactions*, PM21–26.

Hardy, Jean. 2019. "How the Design of Social Technology Fails Rural America." In *Companion Publication of the 2019 on Designing Interactive Systems Conference.* Association for Computing Machinery. https://doi.org/10.1145/3301019.3323906.

Hardy, Jean, and Silvia Lindtner. 2017. "Constructing a Desiring User: Discourse, Rurality, and Design in Location-Based Social Networks." In *Proceedings of the 2017 ACM Conference on Computer Supported Cooperative Work and Social Computing.* Association for Computing Machinery. https://doi.org/10.1145/2998181.2998347.

Hardy, Jean, Chanda Phelan, Morgan Vigil-Hayes, Norman Makoto Su, Susan Wyche, and Phoebe Sengers. 2019. "Designing from the Rural." *Interactions* 26 (4): 37–41. https://doi.org/10.1145/3328487.

Kline, Ronald R. 2000. *Consumers in the Country: Technology and Social Change in Rural America.* The Johns Hopkins University Press.

Mann, Keith. 2018. "A Michigan Township Modernizes Sewer Management with Location Technology." *Esri* (blog), February 22. www.esri.com/about/newsroom/blog/ontwa-sewer-management/.

Sengers, Phoebe. 2011. "What I Learned on Change Islands: Reflections on IT and Pace of Life." *Interactions* 18 (2): 40–8. https://doi.org/10.1145/1925820.1925830.

Sengers, Phoebe, Kaiton Williams, and Vera Khovanskaya. 2021. "Speculation and the Design of Development." *Proceedings of the ACM on Human-Computer Interaction* 5 (CSCW1): 121:1–27. https://doi.org/10.1145/3449195.

Steup, Rosemary, Lynn Dombrowski, and Norman Makoto Su. 2019. "Feeding the World with Data: Visions of Data-Driven Farming." In *Proceedings of the 2019 on Designing Interactive Systems Conference.* Association for Computing Machinery. https://doi.org/10.1145/3322276.3322382.

PART 5

Rethinking Smartness

22 Introduction to Rethinking Smartness

KELLY BRONSON

If smartness has been conceived as a bounded entity located in the mind, or often these days, within digital artefacts, the contributions in part 5, "Rethinking Smartness," trouble bounded smartness. In the first essay, Macktoom and Fatima use a series of photographic panels to visualize the impacts of so-called smart buildings – ones erected in the image of the all-glass "world class" modern city – on the street-level experience of Karachi, Pakistan. The panels display the complex entanglement of modern infrastructure and people's everyday experience of climate change; while the sleek smart building may keep residents inside cool, it actually exacerbates warm temperatures outside. The panels also reveal how city residents respond by devising clever (smart?) ways to modify their experience of heat on the street using mechanisms that have nothing to do with shiny new digital things. Using visuals combined with text to unveil the problematic implementation of "smart" interventions, the authors at once reveal the politics of smartification processes more broadly. Smart cities are predominantly designed in the image of the Global North and thus work at odds with, as the authors put it, "cities that are particularly vulnerable to climate change impacts" (245). The street-level efforts to keep cool and vernacular cooling practices visualized by the author's heat images may be less legible as "smart," but ultimately their functionality reveals the need to align technological innovation and city planning with local social and ecological milieu.

Artist and researcher Bruno Moreschi and political scientist Gabriel Pereira share their short film *Future Movement Future – REJECTED*, in which we follow a panel of academics deciding on and ultimately rejecting a proposal for an algorithmic model based on smart cameras that would predict the movement of cars. Their rejection stands for what Moreschi and Pereira call a "Positive History of NO" to smartness. Their provocation invites us not only to rethink what smartness stands for here but also to consider who should decide what smart systems are to be approved.

Carola Moujan makes a similar argument to Macktoom and Fatima's: smartification is not something universally applicable; moreover, there is a politics to assuming so. Moujan begins her essay with the premise that cities "are complex systems, with unexpected properties that cannot be predicted from the sum of their parts"; interactions between smart city components produce what she calls "emergent behaviors" (269), ones that express any city's adaptable capacity and tendency toward spontaneity. Moujan uses art to problematize simplistic notions of "smartness" to reveal what she calls the "opposition between digital smartness and biological smartness," a kind of false dualistic thinking. Moujan gives us the *Forest City* metaphor or "research creation" to expand perspectives on urban smartness; the forest does not separate objects (like technologies) and subjects like animals, instead revealing the relationships and the "plurality of worlds that escapes representation and is not restricted to human agency." The forest disrupts the hegemonic notions of smartness that abound both in practice and in academic literature on smart cities.

Focusing on the home, Vytautas Jankauskas and Claire Glanois also propose a redefinition of smartification from a singular process or entity to a set of interdependencies among machines and animals. These authors also use art – in the form of video footage of domestic or "smart" robots in action; *Unfamiliar Convenient* documents the relationship between a vacuum cleaner and a voice assistant as part of a "wider ecosystem of a home and its inhabitants" (286). Smart things (like sensors) are shown as perceiving and agential – conditioning the inhabitants sharing domestic space. Working against the image of sensing robots and machine intelligence as capable of generating seamless integration (the Internet of Things image) and human control over the environment, their project "advocates for coexistence, where machines are allowed to exhibit their unique behaviors and adapt to the ever-changing dynamics of domestic life" (287). The authors ask us to approach smart machines with curiosity rather than control, and they ask us to ask ourselves what we might learn from these co-inhabitants.

23 Contesting Smartness in an Unequal City

SOHA MACKTOOM AND AQDAS FATIMA

With rapid urbanization across the globe, there has been a rising interest in the development of smart cities, particularly under the impression that they are resource efficient, sustainable, and better governed. Terms such as "sustainable cities," "smart cities," and "smart buildings and infrastructure" have proliferated in the ways our cities are conceptualized and planned, yet the definition of "smart" remains elusive. Whether it refers to city form, infrastructure, or processes in city planning is vague, and it may involve limited localized implementation and a lack of attention towards potential challenges that may arise. The novelty of the idea of a smart city, especially within the developing world, calls for scrutiny of its effectiveness and interactions with a host of political, socioeconomic, ecological, and spatial factors. These factors in turn force us to rethink what it means for spaces to be considered "smart," particularly in areas of the Global South where infrastructure is often unreliable or entirely absent.

This chapter provides an overview of smart city projects in Pakistan, the actors involved, and their competing narratives that have resulted in what we consider as the problematic implementation of "smart" interventions. We especially focus on the replication of world-class aesthetics under the guise of "smartness" that have long-standing implications for cities that are particularly vulnerable to climate change impacts. In what follows, we begin by discussing the various conceptualizations of smart cities in planning discourses and how their implementation in development projects in Pakistan is diluted to only serve as marketing ploys and visuals of extravagance. We use Karachi, Pakistan's largest city, as a proxy to understand discourses of smart development and the ways in which they link with challenges posed by extreme rises in temperature. We articulate our argument through an art panel (see fig. 23.1), breaking down Karachi's dense public spaces into three crucial elements

Figure 23.1. "Contesting Smartness in an Unequal City." Panel artwork by Soha Macktoom and Aqdas Fatima.

that contribute to ongoing conversations on heating cities: the "smart" built environment, the "unsmart" informal space, and the disintegrated greening practices that are common to both. Using these visuals, we seek to expand the conversation on the impacts of climate change beyond statistical measures to show variegated thermal experiences faced by urban residents. In doing so, we present multi-scalar perspectives on "smartness" in relation to cooling, both in terms of top-down planning and vernacular practices. Subsequently, we propose a need to rethink smartness beyond ubiquitous technologies to one that integrates embedded knowledges and effective practices that are truly representative of smarter environments.

Conceptualizing Smart Cities in Planning Discourses

City planning as a global practice is moving towards incorporating notions of "smartness," with digital technologies converting cities into automated, self-learning systems. Yet, the compounded sociocultural impacts of such transformations remain unexamined, particularly in the ways in which people interact with spaces and environments undergoing technological changes. Planning entails long-term and multi-scalar decision-making, putting it at the centre of establishing any relationship between technology, society, and cities beyond quick technological fixes and cosmetic interventions. While the need to align technological innovation with the built environment is integral to planning future cities

(Karvonen et al. 2020), there is little conversation over whose interests they serve and their potential consequences. Thus, alongside understanding the implications of smart cities on architectural developments and urban planning, an anthropological inquiry into the social, political, cultural, and environmental concerns that emerge as a result becomes necessary. Urban anthropologists such as Lanzeni (2020) and Zimmerman (2018) have written extensively about future imaginaries of smart urban design and reassessed notions of liveability under the pretences of automated cities. Such inquiries by anthropologists bring forth the ways in which metaphors of the "smart city" have complicated the planning and design of cities (Mattern 2021) and thus suggest the need for re-examining notions of smartness.

In the postcolonial context of Pakistan, urban experiences of "smartification" demonstrate a blatant disregard for crucial but less visible dimensions of "smartness," such as wiser management of natural resources, alternative energy systems, operational efficiencies, and participatory governance. These are not solely techno-centric but remain attentive to contextual needs and constraints while contributing to improvements in living arrangements and built environments for urban populations. Instead, what we observe in the country's smart city projects are interventions based on preconceived aesthetics, with glass-cladded high-rise buildings as signifiers of smart development, pursued to boost the city's economic image and business identity (Anwar 2014). These features are characteristic of visions of the "world-class city," which is depicted as a clean, safe city with gleaming skyscrapers that mask socio-spatial inequalities under notions of development (Roy 2014). In the following sections, we problematize the integration of these developmental aesthetics for cities in Pakistan, particularly because they are borrowed from places where political, sociocultural, economic, and ecological contexts are vastly different. We underscore the futility of attempting to address Pakistan's urban challenges through such ill-suited developmental strategies that conflate "smartness" with ideas of the world-class city and simultaneously ignore realities on the ground.

This is especially pertinent for cities in the Global South as they are highly unequal, and these forms of inequality are further exacerbated by impacts of climate change (Masson-Delmotte 2021), which have come to the forefront in conversations about urban challenges. According to the Global Climate Risk Index 2025 Pakistan is the most vulnerable country in the world (Adil et al. 2025), while Sindh – the province within which Karachi is located – is understood as one of South Asia's hotspots for climate change (Mani 2018, 10). This has important implications for ideas

Figure 23.2. Karachi's rapid urbanization.

of the smart city and necessitates bringing ecological, environmental, and anthropogenic concerns to the fore in future planning endeavours, especially now that many regions are experiencing temperature thresholds that are considered biologically unliveable (Mearns and Norton 2009). Using Karachi (see fig. 23.2) enables an exploration of contrasting climate adaptation experiences in cities of the Global South that are informed by specific sociopolitical and ecological contexts, as well as a further examination of discourses of smart development.

Our engagement in a three-year research project titled "Cool Infrastructures: Life with Heat in the Off-Grid City" led us to examine the unique positionality of the city, its complex layers of urban governance and land ownership, and the formal and informal ways in which the city is adapting to the unequal rise in temperatures in its built and social environments. While the project's exploration of exposure and adaptation to heat across cities in South Asia, Southeast Asia, and West Africa revealed similarities within forms of urban development across these cities, it also brought forward Karachi's distinct urban context in Pakistan. This prompted us to analyse the landscape of Karachi from the lens of "smart" and "unsmart" cooling technologies that urban residents engage with on a day-to-day basis in order to rethink the ways "smartness" is understood in cities that are facing the brunt of climate change impacts.

Smart Cities in the Context of Pakistan

The concept of the smart city remains a novel idea in Pakistan's context. Pakistan, a country with a total population of 180 million, has seen its urban population grow drastically in the last few decades (PBS 2023). At present, Pakistan has nine cities with populations over 1 million, all with unique challenges that range from transportation to security and governance, with a lack of resources to support infrastructural development of the health, education, and environment sectors, among others.

In 2019, Capital Smart City Project, Islamabad, was launched by a private real estate development agency. It was advertised as the first "smart city" in Pakistan (fourth in Asia), spread over 40.8 km². With the tagline "building the city of the future," the project's marketing focuses on its many infrastructural and entertainment attractions. However, the "smart" features listed in the project brochure can be argued as those necessary for any functional city, such as public transport facilities, automated traffic control, and a consistent supply of water and energy. Moreover, as we previously discussed, aspirations of the smart "city of the future" in the context of this project bear a striking resemblance to visions of the "world-class city." The visuals available on the project website are spitting images of buildings and locations in Dubai that have been regurgitated time and again to concretize ideas of desirable cities. Despite this facade, the project has paved the way for the development of other smart cities in Pakistan. These include a second smart city project in Lahore initiated by a private real estate developer and military-backed, gated community projects such as DHA (Defence Housing Authority) City phase 9 in Karachi.

We observe that initiatives for smart city projects in Pakistan primarily build two distinct narratives. The first comes from those directly engaged in real estate development projects; for such actors, there is a deliberate attempt to favour new, largely inaccessible gated communities over making existing cities smart as in the projects mentioned above. While the concept of gated communities has already been critiqued in literature (Low 2003), the advent of the "smart" gated community further establishes existing cities as "unfixable" and these communities as the future of urban living. Furthermore, they show little consideration for the complexities of addressing urban challenges through smart interventions. As we discuss in the following sections, similar patterns can be observed in Karachi's development discourse, where these narratives have resulted in the arbitrary implementation of smart interventions.

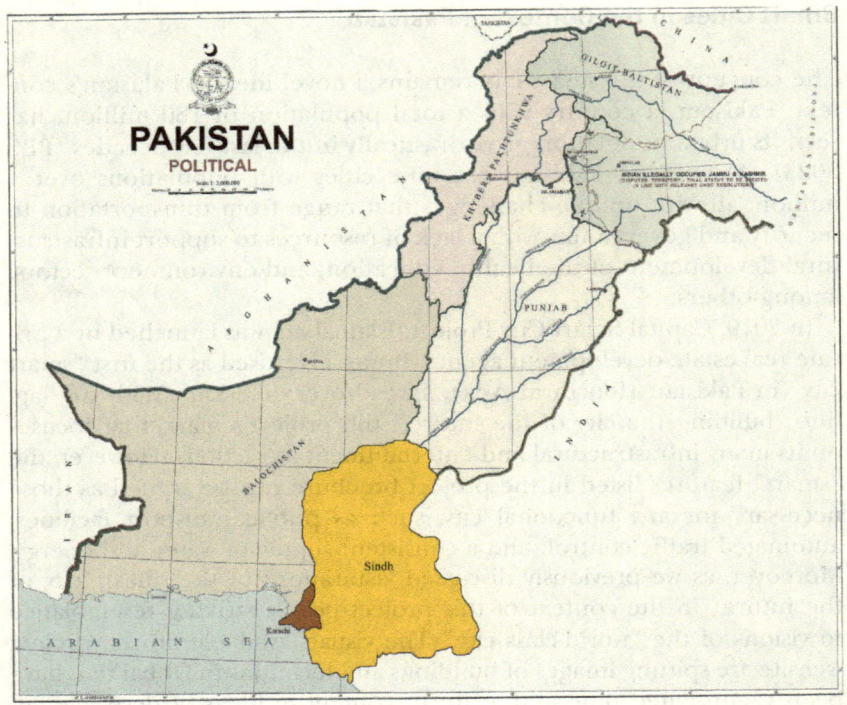

Figure 23.3. Map of Karachi, Pakistan. Source: Political map by Government of Pakistan, annotation by Karachi Urban Lab.

Karachi and the Smart City Narrative

Located along the coast of the Arabian Sea, Karachi is the provincial capital of Sindh, the southernmost province of Pakistan (see fig. 23.3). Karachi's climate is classified as BWh (low altitude, hot, arid, and subtropical climate) under the Koppen-Gieger system (Beck et al. 2018). With an official population of 20.6 million (PBS 2023), Karachi is undeniably indispensable to Pakistan's economy, contributing approximately 25 per cent to the country's GDP and 54 per cent to its tax revenues (see fig. 23.4). A highly unequal city, Karachi's contested access to land, housing, and inequitable infrastructures – in addition to different rates of temperature change in the city's core and peripheral areas – have become significant drivers of people's vulnerabilities (Hasan 2016). Recent scholarship has shown that the daytime temperatures of Karachi have risen by an average of 1.6°C while the night-time temperatures have increased by approximately 2.4°C in the last sixty years, an increase that

Figure 23.4. Karachi's Central Business District.

is substantially higher than the rest of the world (Anwar et al. 2022). Given these alarming statistics, it is crucial to examine Karachi's urban development discourse and future visions of "smartness," especially in relation to heat.

Despite repetitive calls for recognizing Karachi as a smart city by the different stakeholders who lay claim over its decision-making, there is little consensus over what it means to make the city smart or the avenues within which it is most required. These multiple stakeholders range from competing political parties, different political representation at federal, provincial, and local levels, over nineteen land-owning agencies, international unilateral and bilateral organizations such as the World Bank and the Asian Development Bank, and non-state actors including civil and welfare organizations.

Over the last few years, the growing chorus for Karachi's "smartification" has included statements from the city mayor, a representative of the local government, who vocally expressed the progress towards becoming a smart city as one that follows the footsteps of cities like "Singapore, Amsterdam, Barcelona, Madrid and Stockholm" (Dawn 2017). Contrary to this vision, the governor of Sindh, on behalf of the federal government, equated the visions of a smart city with control: "a city like Karachi could not be controlled without becoming a smart city. The sooner we declare Karachi a safe city, the better" (The Nation 2022). While this narrative suggests that "smartness" is a mechanism through which the city can be made safer, the mayor's vision for Karachi relies on the imitation of development aesthetics from elsewhere. Simultaneously, city diagnostic reports by the World Bank aimed towards "Transforming Karachi into a Livable and Competitive Megacity" consider one of the key pillars for city transformation to be "creating a smart Karachi through policies and use of smart tools and technologies" (World Bank 2018). However, this single sentence, used multiple times in the report, is nowhere supplemented by elaborations or guidelines on how this is to be achieved, making the term smart cities nothing more than a buzzword for city improvement.

We posit that such smart city benchmarks are not only climatically ill-suited given the choice of material but also rely on energy-intensive designs that are not feasible in resource-poor countries. Hence, these calls remain isolated from the actual problems faced by everyday residents and instead reimagine the city by reinforcing existing social inequalities and power dynamics. In addition, certain non-state actors working towards improving the efficiency of infrastructural provision through innovative technology regularly go unnoticed in the "smart" landscape of Karachi. For example, the Al-Khidmat Foundation, a non-profit welfare organization, has recently launched "water ATMs" at its water filtration plants in Karachi. The concept of the water ATM (automated teller machine) is similar to that of regular ATMs: people who are registered card holders can simply use them to access water from designated filtration plants. Water ATMs guarantee twenty-four-hour provision of clean drinking water, an initiative that is particularly important in cities such as Karachi, which are moving dangerously close to water scarcity in an increasingly hot climate. However, there has been little conversation over recognizing this initiative as "sustainable smartness," and much of the official government narrative on "smartness" remains tainted with images of shiny high-rises.

Who Is Smart in an Unequal Karachi?

In what follows, we consolidate our observations of top-down as well as vernacular cooling practices found throughout the city of Karachi. We use a combination of photography, thermal imagery, and illustration to visualize the disparities in the effectiveness of these cooling practices and their disproportionate impacts, revealing how access to thermal comfort is unevenly distributed across social and spatial lines. The thermal images (figs. 23.1, 23.5, 23.7, 23.8) display contrasting temperatures on material surfaces using colour gradations (the warmest parts of the image are depicted in yellow and the coolest in blue). Our intention with the use of visual methods is grounded in the realization that conversations over the impacts of climate change are often dominated by wider statistics on temperature changes, thereby overlooking everyday weather and climate experiences (Frazier 2019). In particular, the use of

Figure 23.5. "Smart" building practices in Karachi following visions of the world-class city.

Figure 23.6. Materials on the south sides of buildings with contrasting thermal properties.

thermal imagery reveals the stark contrasts between spaces that are associated with "smart" and "unsmart" cooling technologies. By forming this visual landscape, we examine several cooling practices, materials, and technologies, particularly as they interplay with spatial and sociopolitical inequalities. Collages of these images are superimposed with sketches to contextualize the elements explored within each panel.

Through this series of panels, we break down dense public spaces in Karachi's Southern District into three crucial elements that contribute to ongoing conversations on heating cities: the "smart" built environment, the "unsmart" informal space, and the disintegrated greening practices that are common to both. In the first panel (fig. 23.5), we collect images of the emerging high-rise constructions that utilize materials that are climatically inappropriate in Karachi's hot, arid environment. These high-rises have often been used as poster images to present the "global," "business friendly," and "progressive" image of the city. Buildings such as the MCB Tower, UBL Tower, and IBM Building, most of which have facades largely covered in glass, recorded temperature readings on their surfaces as high as 66.6°C, while buildings constructed with stone blocks displayed temperature readings of 24.3°C. (See fig. 23.6).

These stone buildings demonstrate no distinction between different extents of solar exposure on different facades, reflecting sun rays directly from these shiny glass and metal structures onto those occupying the streets during peak afternoon hours. Other surface materials such as asphalt on the road, metal street furniture, and signage were evidently more heated than surrounding elements such as stone pavements and wood-clad surfaces (fig. 23.6). "Smart" interventions such as solar panels, air conditioning units, cars, and passive electrical devices such as transformers were largely inaccessible by the majority of the public and, in most cases, reflected heat onto surrounding spaces. Most notably, building facades covered in air conditioning outer units emitted hot air onto the surrounding streets, intensifying the heat experienced by pedestrians and occupants of informal markets within the area. These "smart" technologies, introduced to control and manage indoor environments, end up exacerbating the experience of heat for urban residents who do not have access to these technologies.

Figure 23.7, which includes images collected from different formal and informal markets in the Saddar area, counters this notion of "smart" by investigating localized practices often labelled as "unsmart," but which can be observed to manage heat efficiently. Such practices include the use of fabric for shade, clay pots for storage of drinking water, and makeshift stalls that can be reoriented throughout the day, actively mitigating impacts of heat by keeping temperatures between 29°C and 32°C on hot summer days. Informal heat mitigation strategies often work in tandem with the properties of stone used to construct the colonial buildings in which most of such markets are set up. In Karachi's climate, where average day temperatures range between 30°C and 40°C, this – given the insulation properties of these forty-five-centimetre-thick stone blocks – can contribute to indoor temperatures as low as 28°C. In addition, they also provide shaded nooks for several informal vendors, which is especially important considering the lack of other cooling technologies such as fans and air conditioners, which could lower the temperatures to considerably more comfortable levels but at the cost of unaffordable electricity bills. Given these circumstances, the ability of the building materials to manage temperatures in indoor and sometimes in semi-indoor spaces provides a "smart" alternative to otherwise inaccessible technologies. Other practices observed included wall mounted fans installed to keep both the vendor and the vegetable produce in certain markets under consistent ventilation, awnings extending beyond shop boundaries ensuring shade for the customers, and umbrella style sheds at different traffic intersections of the city to provide shade for the traffic policemen who spend their entire day outdoors under the direct sun.

Figure 23.8 depicts the way greening practices remain isolated from the overall understanding of climate management. Imported plantations and practices such as palm trees, roof gardens, and seasonal flowers are

Figure 23.7. "Unsmart" vernacular cooling practices in Karachi's markets.

Figure 23.8. "Smart" (left) and "unsmart" (right) greening practices across Karachi.

characteristic of "smart" landscaping, serving little purpose beyond aesthetic greening. As one of the images shows, temperature contrasts between concrete buildings and palm trees were hardly noticeable. The "smartness" in using palm species in urban outdoor spaces – yet another example of the conflation of smart attributes with "Dubaization" (Elsheshtawy 2012) – is questionable given their lack of shade. In contrast, rapidly disappearing native species such as the Banyan tree are routinely used by pedestrians and informal workers for their shade (fig. 23.9). While efforts to preserve these species are underway, particularly in some old neighbourhoods, as photographed above (fig. 23.8), little has been done to encourage the inclusion of these species in ongoing "plantation drives" across the city. Similarly, mangroves are quickly disappearing under ongoing land reclamation practices by formal and informal, low-income settlements, given the land pressures in a rapidly urbanizing city. This is particularly worrying as these species are characteristic of coastal cities in South Asia and are considered essential for healthy coastal ecosystems.

Conclusion

As we have explored in this chapter, Pakistan's current narratives of infrastructural improvements are geared towards world-class city aspirations under the guise of "smartness." Consequently, they often homogenize discourses of "city" experiences that gloss over uneven terrains of liveability and limit infrastructural development to what is visible, borrowing from western notions of modernity (Anwar 2014). We visualize the unequal city not only through aesthetics of the built environment but also in the ways in which globally informed ideas of smartification interfere with culturally grounded practices. In doing so, we present multi-scalar perspectives on "smartness" when it comes to cooling and conceptualize ways of rethinking smartness and smart practices, particularly in southern cities. Having visualized the disparities of smartness in the unequal city and how these ongoing practices contribute to exacerbating impacts of climate change, we prompt the reimagination of the "smart" Karachi of the future. We underscore that the domains of planning and anthropology are well placed to be at the centre of policy interventions that include intersectionalities of urban development with the bodily experience of heat. Contributions from such disciplines will allow for locally placed understandings of "smartness" that are grounded in vernacular cooling practices. These will be essential in rethinking a "smart" future for Karachi – one that is accessible and effective in shaping stronger relationships between people and their built environments.

Figures 23.9. Activity under "unsmart" green cover versus disintegrated "smart" landscaping.

Funding Acknowledgment

The research for this chapter is an output of a broader three-year project entitled "Cool Infrastructures: Life with Heat in the Off-Grid City," supported by the United Kingdom's Economic and Social Research Council as part of the Global Challenges Research Fund.

References

Adil, L., D. Eckstein, V. Künzel, and L. Schäfer. 2025. "Climate Risk Index 2025." Germanwatch. https://www.germanwatch.org/sites/default/files/2025-02/Climate%20Risk%20Index%202025.pdf.

Anwar, Nausheen H. 2014. "The Postcolonial City in South Asia." *Singapore Journal of Tropical Geography* 35 (1): 22–38. https://doi.org/10.1111/sjtg.12048.

Anwar, Nausheen H., Hassaan F. Khan, Adam Abdullah, Soha Macktoom, and Aqdas Fatima. 2022. *Heat Governance in Urban South Asia: The Case of Karachi*. Cool Infrastructure, UKRI (GCRF). https://doi.org/10.7488/era/2180.

Beck, Hylke E., Niklaus E. Zimmermann, Tim R. McVicar, Noemi Vergopolan, Alexis Berg, and Eric F. Wood. 2018. "Present and Future Köppen-Geiger Climate Classification Maps at 1-km Resolution." *Scientific Data* 5 (1): 1–12. https://doi.org/10.1038/sdata.2018.214.

Dawn. 2017. "Karachi to Be Made Smart City, Says Mayor." *DAWN*, July 4. www.dawn.com/news/1343006.

Elsheshtawy, Yasser. 2012. "Dubaization: Reflections on Middle Eastern Urbanism, Inspired by Dubai." http://dubaization.com/dubaization.

Frazier, Camille. 2019. "Urban Heat: Rising Temperatures as Critique in India's Air-Conditioned City." *City & Society* 31 (3): 441–61. https://doi.org/10.1111/ciso.12228.

Hasan, Arif. 2016. "Emerging Urbanisation Trends: The Case of Karachi. Ref. Number C-37319-PAK-1." Working Paper for the International Growth Center, London School of Economics, London, UK.

Karvonen, Andrew, Matthew Cook, and Håvard Haarstad. 2020. "Urban Planning and the Smart City: Projects, Practices and Politics." *Urban Planning* 5 (1): 65–8. https://doi.org/10.17645/up.v5i1.2936.

Lanzeni, Débora. 2020. "Smart Global Futures: Designing Affordable Materialities for a Better Life." In *Digital Materialities: Design and Anthropology*, edited by Sarah Pink, Elisenda Ardèvol, and Dèbora Lanzeni. Taylor Francis. https://doi.org/10.4324/9781003085218-4.

Low, Setha M. 2003. *Behind the Gates: The New American Dream – Searching for Security in America*. Routledge.

Mani, Muthukumara, Sushenjit Bandyopadhyay, Shun Chonabayashi, and Anil Markandya. 2018. *South Asia's Hotspots: The Impact of Temperature and Precipitation Changes on Living Standards*. World Bank. https://doi.org/10.1596/978-1-4648-1155-5.

Masson-Delmotte, Valérie, Panmao Zhai, Anna Pirani et al. 2021. "Climate Change 2021: The Physical Science Basis." Contribution of Working Group I to the Sixth Assessment Report of the Intergovernmental Panel on

Climate Change. https://www.ipcc.ch/report/ar6/wg1/downloads/report/IPCC_AR6_WGI_SPM_final.pdf.

Mattern, Shannon. 2021. *A City Is Not a Computer: Other Urban Intelligences.* Princeton University Press. https://doi.org/10.1515/9780691226750.

Mearns, Robin, and Andrew Norton, eds. 2009. *Social Dimensions of Climate Change: Equity and Vulnerability in a Warming World.* World Bank. https://doi.org/10.1596/978-0-8213-7887-8.

PBS. 2000. "District Census Report of Karachi South, Pakistan." Demographic Survey 1998, Pakistan Bureau of Statistics, Statistics Division, Government of Pakistan, Islamabad. http://digitalarchive.uet.edu.pk/handle/123456789/583.

— 2023. "Table 1: Area, Population by Sex, Sex Ratio, Population Density, Urban Population, Household Size and Annual Growth Rate, Census-2023 Sindh." https://www.pbs.gov.pk/sites/default/files/population/2023/tables/table_1_sindh_districts.pdf.

Roy, Ananya. 2014. "Slum-free Cities of the Asian Century: Postcolonial Government and the Project of Inclusive Growth." *Singapore Journal of Tropical Geography* 35 (1): 136–50. https://doi.org/10.1111/sjtg.12047.

The Nation. 2022. "Karachi Must Be Declared Smart City, Says Sindh Governor." *The Nation*, January 14. www.nation.com.pk/14-Jan-2022/karachi-must-be-declared-smart-city-says-sindh-governor.

World Bank. 2018. *Transforming Karachi into a Livable and Competitive Megacity: A City Diagnostic and Transformation Strategy.* The World Bank. https://hdl.handle.net/10986/29376.

Zimmerman, Katrina Johnston. 2018. "Consider Anthropology in Your Next Urban Design Project." *Meetings of the Mind*, March 6. https://web.archive.org/web/20231130015423/https://meetingoftheminds.org/consider-anthropology-next-urban-design-project-25264.

24 Future Movement
Future – REJECTED

BRUNO MORESCHI AND GABRIEL PEREIRA

Artists' Statement

Perhaps the time has come to collectively build a History of Smartness that does not only include acceptances, advances, and challenges overcome. Rather, it should include key moments in which NO was the smartest, most sensible, and fundamental attitude. This Positive History of NO should not be confused with technophobic or authoritarian denials. It is time we refuse the hype, move slow, and refuse the problematic potentials of smart technologies.

A compendium of smart NOs may begin with the history of the German philosopher and mathematician Gottfried Wilhelm Leibniz, known for perfecting the binary numbers that would form the basis of today's computers. For years, Leibniz attempted to gain an audience with French Emperor Louis XIV to convince him to invade Egypt, thus directing the Emperor's armies away from European conquest. Leibniz presented a compelling case: hundreds of route maps, tables with the exact number of needed armies, the order of which cities to attack, a plan B and a plan C to deal with the unexpected, etc. For months, perhaps years, Leibniz had been in Paris trying to convince Le Roi Soleil's ministers to arrange a meeting at the palace in Versailles. Historical accounts suggest that the king's ministers were stalling for time. Because of that, a meeting never actually took place. The ministers were convinced that past endeavours were stronger than graphic, cartographic, and statistical materials in the present. It had been four centuries since France had embarked on a crusade – why do something similar at that moment? Lucky for Egypt, for the rest of Europe, for the whole world.

The advancement of contemporary technology also needs refusals to lead to truly smart advances. Borrowing from science fiction and meme aesthetics, our speculative contribution to a Positive History of NO is the short film "Future Movement Future – REJECTED." In the film, we listen to academics review a proposal for funding a new algorithmic model for smart cameras which analyses the movement of cars on a never-before-seen scale. This system claims to allow the emergence of new forms of predictive capabilities: its algorithm would be able, for example, to predict the routes that drivers want to take, even before they take them. But the panel of researchers from three different universities audit and emphatically reject the proposed system.

At least in this case, the future of dystopian surveillance was barred by institutional refusal in an attitude that shows that smartness is also knowing precisely what should (not) be approved and encouraged. However, saying NO comes with its own questions: who are the people in this panel? Why are they given the mandate to refuse the project? Should we trust their analysis? Like never before, it is time for thinking not just about what forms of smartness and smartification processes we (do not) want, but also about how we want to make these decisions.

The short film was born in a moment when much discussion was emerging on the immense problems that ubiquitous "smart" solutions were generating, particularly for marginalized communities. A growing sentiment is that not only are these systems being uncritically woven into our lives, but they are becoming the default we must accept. The only possible way out is to say NO. As the scholar Seeta Peña Gangadharan has phrased it, people "are unwilling to accept data-driven systems in the terms and conditions that government or private actors present to us" (2019, 3).

It is up to us all to strike back by finding ways to refuse algorithmic control through a politics of antagonistic and disruptive tactical action (Heemsbergen et al. 2022). An example of this is the growing fascination of researchers, artists, and activists with the Luddites. As analysed by media scholar Gavin Mueller (2021), these English textile workers, already two centuries ago, collectively revolted against the automation of their labour that was taking away their power and rights. Mueller sees this deep history as a possible frame for thinking of a decelerationist project, forming a new class to challenge the dominant frame of contemporary digital capitalism.

Luddism is, in this sense, not only against the efficiency of production, but also an antagonistic stance: "it sets itself against existing capitalist social relations, which can only end through struggle, not through factors like state reforms, the increasing superfluity of goods, or a better planned economy" (Mueller 2021, 129). A decelerationist project thus refuses appropriation by standing outside the technological solutionism of current times, proposing an alternative to the uncontrolled expansion

Screenshot 1 of short film *Future Movement Future – REJECTED*

of "smartness." As summarized by political economist Jathan Sadowski (2021), such antagonisms would mean confronting "the harms done by digital capitalism and seek[ing] to address them by giving people more power over the technological systems that structure their lives" (see also Velkova and Kaun 2021).

But what does such control look like, particularly considering the unbalanced power relations of current data regimes? How could collective refusal take shape in practice, and what kinds of institutions could support these practices? Who, in fact, has the right to refuse smartness? Tentative responses have been cropping up. Cifor et al. (2019) have proposed a "Feminist Data Manifest-No," a declaration of ten points of refusal to current harmful data regimes and commitment to new data futures. Among those is: "We refuse the expansion of forms of data science that normalizes a condition of data extractivism and is defined primarily by the drive to monetize and hyper-individualize the human experience" (2019). The work of activists and researchers has become more than ever "dismantling carceral technologies" (Hamid 2020), fighting back alongside affected communities (see Tierra Comun Network 2023), and enacting practices that refuse the use of data for surveillance (Pereira 2021).

As can be seen, saying NO is much more than a negative statement. It can serve as a generative approach, helping to rethink our approach to "smartness" and considering how things could be otherwise – as hard as that may be.

The future of smartness is up for grabs. It is time we learn from the Positive History of NO.

Screenshot 2 of short film *Future Movement Future – REJECTED*

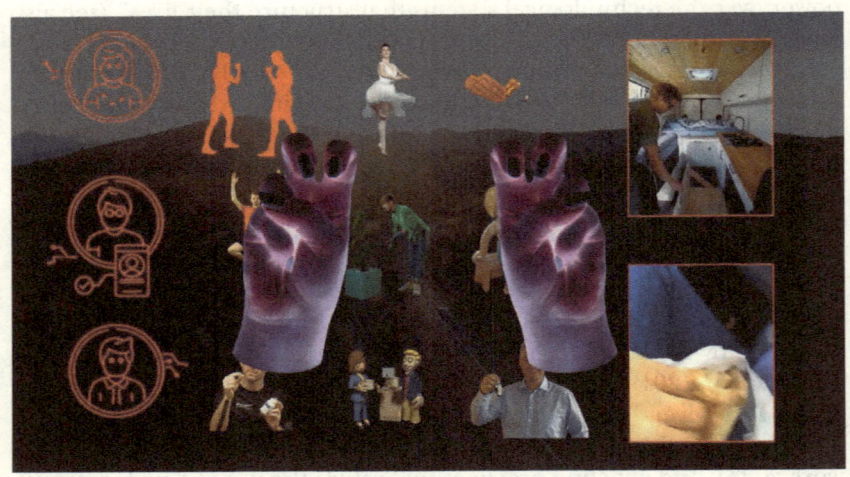

Screenshot 3 of short film *Future Movement Future – REJECTED*

Future Movement Future – REJECTED can be watched at https://vimeo.com/654303920.

References

Cifor, M., P. Garcia, T.L. Cowan, J. Rault, T. Sutherland, A. Chan, J. Rode, A.L. Hoffmann, N. Salehi, and L. Nakamura. 2019. "Feminist Data Manifest-No." www.manifestno.com/.

Gangadharan, Seeta Peña, dir. 2019. "Technologies of Control and Our Right of Refusal." TEDx. http://eprints.lse.ac.uk/101157/4/Gangadharan_technologies_of_control_remarks.pdf.

Hamid, Sarah T. 2020. "Community Defense: Sarah T. Hamid on Abolishing Carceral Technologies." *Logic Magazine*, no. 11 (August). https://logicmag.io/care/community-defense-sarah-t-hamid-on-abolishing-carceral-technologies/.

Heemsbergen, Luke, Emiliano Treré, and Gabriel Pereira. 2022. "Introduction to Algorithmic Antagonisms: Resistance, Reconfiguration, and Renaissance for Computational Life." *Media International Australia* 183 (1): 3–15. https://doi.org/10.1177/1329878X221086042.

Mueller, Gavin. 2021. *Breaking Things at Work: The Luddites Are Right About Why You Hate Your Job*. Verso.

Pereira, Gabriel. 2021. "Towards Refusing as a Critical Technical Practice: Struggling with Hegemonic Computer Vision." *A Peer-Reviewed Journal About Research Refusal* 10 (1): 30–43. https://doi.org/10.7146/aprja.v10i1.128185.

Sadowski, Jathan. 2021. "I'm a Luddite. You Should Be One Too." *The Conversation*. http://theconversation.com/im-a-luddite-you-should-be-one-too-163172.

The Tierra Comun Network, ed. 2023. *Resisting Data Colonialism: A Practical Intervention*. Institute of Network Cultures.

Velkova, Julia, and Anne Kaun. 2021. "Algorithmic Resistance: Media Practices and the Politics of Repair." *Information, Communication & Society* 24 (4): 523–40. https://doi.org/10.1080/1369118X.2019.1657162.

25 Thinking like a City

CAROLA MOUJAN

Introduction

Cities are complex systems, with unexpected properties that cannot be predicted from the sum of their parts. Interactions between components, actors, and processes produce emergent behaviours that change the expected course of actions, expressing the city's capacity to adapt to unforeseeable conditions and achieve spontaneous order (Ladyman et al. 2013). In that sense, urban smartness can be seen as the multiplicity of forms of self-agency that express urban life. Yet, because the Smart City paradigm was coined in the digital world, referring to anything as being "smart" immediately suggests embedded digital components, as if smartness was contained in technology. But, of course, urban smartness did not begin with digital technologies: cities are shaped by the interplay of biological, social, affective, material, and technical relationships that give places their unique identities (Rykwert 2002). Moreover, technocentred models of smartness have failed to respond to the magnitude of current problems: most Smart City projects are environmentally harmful, produce unwanted effects, lead to social injustice, and are a threat to people's freedom (Jankowski et al. 2023). The best-known illustration may be the South Korean city of Songdo, where the multiplication of digital services has led to massive installations of surveillance cameras, while many smart services do not perform as expected because their infrastructure model is too rigid to adapt to the messiness and unpredictability of real places. The pneumatic waste collection system below the streets might be the most extreme example: connected to every building, it was designed to eliminate the need for collection trucks. At street level, however, rubbish bins must be activated using an official resident card, which makes them inaccessible to non-residents and often even to residents themselves, as most people forget their cards and rubbish ends up accumulating on sidewalks. "The urban system in Songdo was not flexible and formulated as a closed system which underappreciated the impact of fluctuations and variability ... the closed system approach was constrained by a static conception of the role of foreign/local actors ... a technologically deterministic approach to achieving efficiency ... less sustainable in its ability to cope with complexity" (Mullins 2017).

Another observable trend is a perceived opposition between digital smartness and biological smartness, in the form of dualistic thinking that only replaces one paradigm with the other without challenging the reasoning itself. Indeed, the temptation of "getting back to the wrong nature" (Cronon 1996; Haraway 1988), to a "natural" state of the city driven by organic forces alone, offers no viable alternative. There is no such thing as a "natural" city: technology and city-making are closely intertwined from day one (Mumford [1934] 2010). Moreover, technical and social evolution are enmeshed in multiple ways beyond prescribed uses and appointed ends, and any bio-inspired strategy presents the risk of substituting technical substantialism with reified forms of vitalism that re-entangle us in static forms of reasoning.

To rise to the major challenges contemporary urban environments face, then, requires the "ethico-political articulation between the three ecological registers (the environment, social relations, and human subjectivity)," which Félix Guattari has named *ecosophy*; an "authentic political, social and cultural revolution" (Guattari [1989] 2000, 28). A revolution does not just change ways of *doing*: it transforms ways of *thinking*. What is needed are radically new forms of partnership with more-than-human forces that, instead of controlling, make room for their power and agency; new processes, methodologies, and principles that engage with digital technologies in open and unexpected ways; and a radical shift in reasoning, from modern rationality that focuses primarily on isolated entities, to new "eco-logical" forms of thinking that concentrate on relationships. To look not only at "visible relations of force at a grand scale" but also consider "molecular domains of sensibility, intelligence, and desire" (Guattari [1989] 2000, 28). Weaving research, theory, and art, this chapter aims to rethink smartness and smartification processes from a situated and historically grounded perspective, looking for new ways of connecting technical systems with natural processes and principles, through "denaturalized understandings of the ecological paradigm that emphasize form, time, and dynamic complexity" (Hörl 2017, 33).

I contend that the problem with techno-centred practices of smartification lies in the territorialization of technology, rather than in technology itself. The primacy of screens, the smartphone as a universal interface, and the myth that performance increases with technical complexity are all cultural constructs that can, and should, be challenged. Artists and hackers, for instance, dismantle and rearrange parts of existing technical systems and processes using inexpensive components and DIY techniques, playing with errors and accidents and bending their logic beyond its intended purposes (de Lange and de Waal 2019, 2). One example is the *Google Maps Hacks* (2020), a performance by Berlin-based artist Simon Weckert, in which a handcart full of ninety-nine second-hand phones misguide algorithms to give the impression of a huge traffic jam. Such countercultural assemblages explore situated strategies that take full account of the entangled nature of urban life and work within it, revealing potential uses that could not be imagined from mainstream perspectives (thus expanding the realm of technical possibilities). This is to say that to tinker with technical objects, cobble them together, hack them, and make them deviate from their original purposes is an essential part of what makes them evolve from an initial, "abstract" state of conceptual simplicity where each component performs a dedicated function, to a "concrete" one, where relationships between components support multiple synergies at different levels and systems show emerging properties that are more than the sum of their parts (Simondon [1958] 2017). It is in this intrinsic, autopoietic sense, that a technical object can be considered "smart."

Reflecting on three of my own research-creation projects, I will suggest an alternative approach to urban smartness that I will call *Forest City*. The concept builds upon Eduardo Kohn's definition of a forest as a wealth of semiotic processes connecting human and non-human agents rather than as a physical location or a group of trees (Kohn 2013). Following C.S. Pierce, the author distinguishes three broad classes of signs involved in forest-thinking processes: icons, indices, and symbols. Each class is characterized by the relationship it maintains with the entity it represents: icons convey meaning through similarity, indices point to actual events, and symbols – a class specific to humankind – refer to conventional relationships between signs, understandable only within a given context, like words in human languages. Kohn provides detailed accounts of the processes through which human and non-human inhabitants of the Peruvian rainforest become aware of each other and develop forms of interaction that are not language based.

The *Forest City* metaphor is useful to expand perspectives on urban smartness because it shifts the focus from separate "objects" and "subjects" (the tree, the animal, the human, technology, etc.), to the relationships, and the processes through which relationships emerge and evolve, displacing the primacy of language over other signs, thus opening up a plurality of worlds that escapes representation and is not restricted to human agency. Indeed, beyond the symbolic register distinctive of human semiotics, forests think in images, and images, Kohn (2013) argues, are what we have in common with other species. A *Forest City* vision of urban smartness could be achieved by shifting attention from substances to processes of becoming, where the three eco-logical registers – the environment, social relations, and human subjectivity – connect. Becoming is a movement that deterritorializes elements from the relationships defining them and reassembles them through new, "partical" connections (Viveiros de Castro [2009] 2014). To shift the focus to relationships, then, means attending to specific ways through which technical systems become machines of subjectivation. It is in this sense that we can speak of deterritorialization: through practices that engage with technical systems in creative, open-ended ways not driven by solutionist aims, reconnecting them with living processes and molecular domains of subjectivity.

Methodological Approach

I am an artist, designer, and researcher, with a background in architecture and visual arts, working professionally with digital technologies for three decades. Grounded in creative practice, influenced by the work of Deleuze and Guattari and alternative streams of thought such as early British cybernetics (Pickering 2009), techno-feminism (Haraway 1991), and hacker culture (de Lange and de Waal 2019), my work posits research-creation as a powerful critical tool to deterritorialize hegemonic visions of digital smartness. Defined as the complex intersection of art practice, theoretical concepts, and research, research-creation differs from other forms of scholarship and practice in the way it engages the three components. Artworks are not illustrations of theoretical concepts, nor is the role of theory to explain the artworks. Instead, both work as stepping stones in conceptual development. Unlike more traditional methodologies, research-creation does not seek to solve predefined problems and questions but produces concepts that problematize: questions are not known in advance (Springgay and Truman 2018). Such a process-based critique goes beyond narrative to act upon situations and sketch other futures (Pickering 2009) through bio-techno-social assemblages that follow the flow of materials and affects.

My strategy relies on a critique of the normative role of mainstream technical systems, through uncanny assemblages and situations that deterritorialize technology, suggesting other eco-logical possibilities. In the following section, I will draw a line connecting three of my research-creation projects with current debates around urban smartness, highlighting the relational space each project defines and how it seeks to connect digital smartness with domains of sensibility, intelligence, and desire (Guattari [1989] 2000). The first one, *Luciole*, uses light, datasets, and Internet of Things (IoT) networks to catalyse urban atmospheres and social interactions through soft light modulations. Light patterns and variations reveal the enmeshment of activities, affects, and social interactions at the core of urban places. *[RIP]_Montevideo*, the second one, explores non-linear temporalities and the ambivalence of any heritage preservation strategy. The multiplicity of time frames embedded in places and the active role citizens play in shaping urban memory through inhabitation are made visible through an experimental interface that relies on a destructive gesture. Finally, in *Future Forest Diorama*,[1] the third project, cooperation and mutual help principles from botanical communities serve as the basis for an interactive installation concept where a network of smart terrariums interact with humans using sensors, IoT, and prototyping boards.

Made of low-tech components, open-source technologies, and DIY tools, the projects stage a wealth of relationships as interlinked and complex as those of forests. While only the last one engages directly with vegetation, all can be seen as examples of *Forest City*. Indeed: the *Forest City* metaphor is not a literal reference to forests or trees: a city or place can be *Forest City* without directly involving plants, instead connecting elements like forests do. Addressing this concept only through works that make explicit reference to the botanical world would be too literal and would work against my argument.

My goal is twofold: on the one hand, I seek to challenge conventional notions of science, technology, and art through research-creation, showing how such practices can help avoid dualistic thinking and the hegemony of language. On the other hand, I want to highlight what I consider the most promising potential of digital smartness: the capacity to connect worlds of radical alterity, addressing, exploring, and expanding some of these complex entwinements through an open-ended approach.

Machines of Subjectivation

Let us take a closer at *Luciole*, a network of interactive benches designed for Lyon City Design Urban Forum in 2015 in France. The challenge description focused on transition areas generated by renovation works

---

[1] Developed in collaboration with Agustin Ortiz Herrera.

Figure 25.1. *Luciole* at Lyon City Design Urban Forum, Lyon, 2015.

to ask how such no-go places could become living spaces and create engagement with citizens during construction. Taking the public bench as a starting point, I asked myself whether digitally augmented benches could do even more for city life than their analogue counterparts. The idea of using networks and IoT technologies to activate spaces, trigger playful social interactions, and create local synergies immediately came to mind, as did that of situating mobility-related datasets in the benches without relying on screens. Avoiding screens was the key to addressing common issues, such as the privatization of public spaces, surveillance, privacy threats, high energy consumption, and lack of inclusivity. Indeed, without large screens, there is no advertising; without smartphones, no individual tracking of movements and no need to have specific equipment or knowledge to benefit from the system.

Luciole benches are akin to luminous bubbles made of a metal mesh with embedded optic fibre that use light and short text messages on low-tech displays to convey meaning. The light modulations of *Luciole* transform atmospheres while providing basic functional services that allow users to enjoy the simple pleasure of "being there" without having to focus on screens (Moujan 2015). The benches monitor surrounding activity through sensors, detect sitting, and communicate with the other benches through light diffused by the optic fibre. At the bench level,

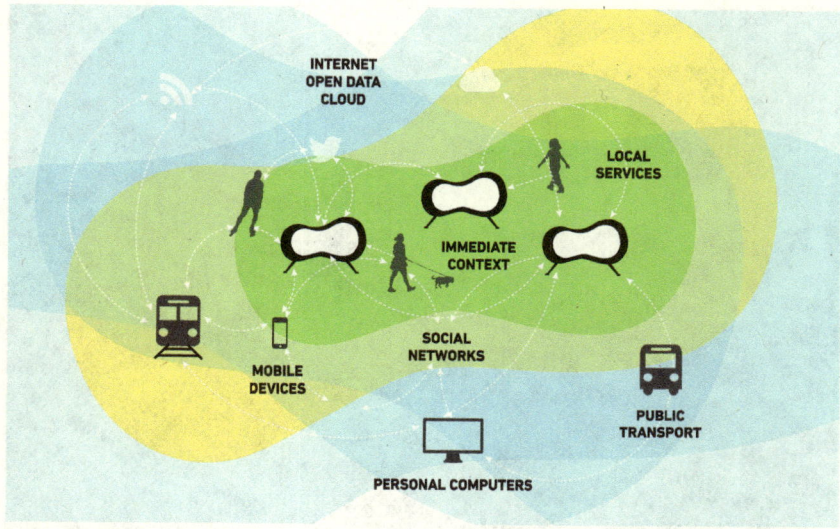

Figure 25.2. Conceptual diagram.

aside from providing seating space, they display real-time information on surrounding activities and services. At the network level, the luminous modulations make visible the flows and presence on site, enlivening degraded places and conveying feelings of serenity.

Implementing the simple functional principles of *Luciole* did not involve overcoming major technical barriers. Yet, building the prototype was challenging because instead of rearranging existing technical bricks, it required assembling parts from different systems. Everything, from hardware to software, needed to be designed, a more complex task involving many hours of qualified work. The process revealed how difficult it is to introduce techno-diversity within public space, how much control there is over public infrastructures, and how long the process of deconstructing screen-based representations of digital technology may be (Moujan 2015).

The second example, *[RIP]_Montevideo*, is an interactive installation about the complexity of urban memory and the paradoxes of any heritage conservation initiative. The piece consists of an interactive book, a closed-circuit video projection, an algorithm, an image database of urban things, and two screens facing each other. The book's pages are made of precut fragments that participants are invited to rip off. When they do, electric contact is interrupted, triggering the display of pairs of

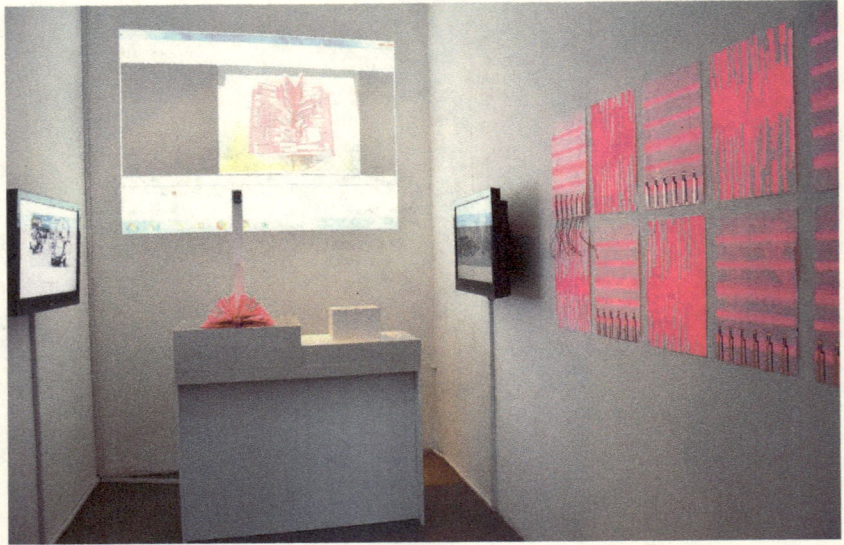

Figure 25.3. *[RIP]_Montevideo*. View of the installation at Espacio de Arte Contemporáneo (EAC).

images on the facing screens, one of which shows an urban object from the past, and the other a Google Street View screenshot of what can be found today at the same place.

The first work-in-process version of *[RIP]_Montevideo* was presented in 2018 at Montevideo's Espacio de Arte Contemporáneo (EAC), a contemporary art centre located in a former prison. The spatial design seeks to amplify affective responses to images of lost urban things through form, materials, gestures, digital interactions, and scenography.

My goal was to highlight the impact of individual choices in the emergence of urban forms. The installation's concept leaves little room for nostalgic contemplation. Violence plays a key role here: the book must be destroyed for the pictures to be seen, and, as participants rip off the pages, the images confront them with the intrinsic ambivalence – between preservation and renewal – of heritage policies. Images are not curated and appear in random order; what counts most is the gesture of destruction performed by the participants themselves (fig. 25.4). The confrontation is further enhanced by the distinctive cell shape of the exhibition space, a constant reminder of the building's previous repressive functions (Moujan 2023).

Once again, building the installation required several iterations. Instead of ripping, many other triggers could have been used, and from

Thinking like a City

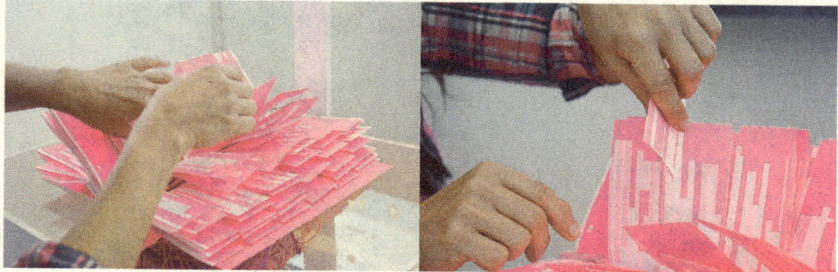

Figure 25.4. Interacting with the book.

a strictly functional perspective, pushing buttons, scanning QR codes, or touching screens would have been much easier paths. The gesture was at odds with technical efficiency, yet its symbolic meaning supported the specific process of sense-making I was aiming for.

Let us now return to the forest/city relationship. Beyond the ornamental and environmental services vegetation provides to cities, plants also play a substantial (and often violent) role in shaping urban forms. Wherever something breaks, a plant springs, and if humans don't interfere, the forest returns, like the tree growing through the roof of an abandoned textile factory shown in figure 25.5. Instead of building green areas and spaces from scratch following traditional composition principles, designers are now beginning to work from and with the vegetation that is already there, more likely to adapt and resist under changing conditions. This means that notions of beauty are also changing along the way. The *Forest City* concept, therefore, refers not only to the need for an increased number of trees and plants within cities but to more complex entwinements between the two entities, new spatiotemporal alliances where, unlike what happens with the rubbish bins of Songdo, roles and agency are not static. A form of collective intelligence, described by Kohn as "living thought" that emerges from differences, including those that may seem like failures when looked at from a technocentric perspective.

"Thoughts and lives grow by capturing difference in the world," writes the author, following Gregory Bateson (Kohn 2013). This brings us to the last example, a research residency involving artistic and scientific partners[2] carried out with Agustín Ortiz Herrera between June 2021 and

---

2 CREAF, a major Catalonian ecology and forestry research centre, and La Escocesa, an artistic research and production centre situated in another one of the cities' former textile factories.

Figure 25.5. Abandoned textile factory in Poblenou, Barcelona, 2020.

March 2022 at the Can Balasc experimental station near Barcelona. We conducted exploratory research necessary to design *Future Forest Diorama*, a speculative installation project consisting of a network of three smart terrariums built inside discarded home appliances (fig. 25.5). Akin to botanical communities, our terrariums would react to surrounding conditions, communicate with each other through IoT networks, and develop survival strategies inspired by continuous relational processes characteristic of "forest thinking" (Kohn 2013). A form of intelligence

Figure 25.6. *Future Forest Diorama* – concept images, 2021.

not defined only by the agents (plants, animals, minerals, humans, fungi, technical systems, tools …) that are part of today's forests but also by the processes through which all those entities interact and inhabit them.

Through this experiment, we wanted to challenge the positivistic principles upon which modern sciences and cultural values have developed: the paradigm of competition as the main driver for biological evolution, and the bucolic images of forests that sustain an illusion of natural purity, reinforcing the "wrong nature" image I stood against at the beginning of the chapter. Deterritorializing the idea of a forest involved exploring and deconstructing it from different perspectives, using a mix of methods and tools coming from the arts, botanical science, and the humanities. Although language was involved, the experiments were image based: we built herbariums, collected and germinated seeds, produced diagrams, photographs, cyanotypes, installations, and walks.

It is in this sense that the Forest City paradigm can be transformative. The metaphor invites us to explore new semiotic processes where urban smartness, forest intelligence, and computational logic may connect: to think more with and through images, to take them seriously, and to listen to what they tell us. During collaborative workshops involving artists and scientists, our images served as tools for interdisciplinary dialogue (Moujan and Ortiz Herrera 2022). This is to say, that images could provide the nexus through which forest-like types of smartness may emerge. Escaping from the hegemony of language is essential to replace human thinking within the continuity of iconic eco-logics shared by all living beings. "Thinking with images … and learning to attend to the ways in which these images amplify, and thus render apparent, something about the human … is also a way of opening ourselves to the distinctive iconic logics of how the forest's thoughts might think their ways through us" (Kohn 2013, 222).

Figure 25.7. *Future Forest Diorama* – the herbarium, 2021.

Toward *Forest City*

While much effort and many resources are directed today toward imagining new ways to combine, implement, and monetize technological building blocks as they are, I have argued that a more fundamental change is required to reconnect digital technologies with the meaningful affordances, forms, rhythms, and affects at the roots of urban smartness. An essential first step is to break away from dominant representations of what technology can do and how it should evolve, that is, to deterritorialize it through new machines of subjectivation. Deterritorialization is a more radical gesture than simply using technical objects for more virtuous ends; to deterritorialize, one needs to intervene in its becoming, polarize force lines in directions that reconfigure associated milieus into new techno-bio-social wholes. The three research-creation projects described in this chapter show how deterritorialization brings about new problems and questions not known in advance, revealing the central role played by concrete prototypes and artworks: that of *spaces of problematization* where initial technical, cultural, and biological frames are displaced. In both *Luciole* and *[RIP]_Montevideo*, technical interventions that appeared conceptually simple became complex because their implementation required redefining their associated technical milieus (Simondon [1958] 2017). In *Future Forest Diorama*, confronting artistic and scientific views revealed hidden dimensions of forests, such as the wealth of cultural values and affects (for the scientists) and the complex biological processes and time frames that sustain them (for the artists), transforming the very way each group sees forests and cities (Moujan and Ortiz Herrera 2022).

Setting the stage for forest-like smartness in cities demands overcoming positivistic approaches based on isolation and control, to allow regenerative, spontaneous, auto-organized processes to emerge: "an arrangement of forces and intensities at once spatial, material and ethological, and constituted by heterogeneous forms of practice that include relations, if not alliances, with other-than-human company" (Barua and Sinha 2022). Such processes differ from traditional organicist models where each part plays a dedicated role, staging new types of part-whole relations. They have the agency of *assemblages*, with emergent properties – "the ability to make something happen" – that come from their "uneven topography" (Bennett 2010). They are "open-ended collectives" with distinctive lifelines. The elements of these assemblages, "while they include humans and their (social, legal, linguistic) constructions, also include some very powerful nonhumans: electrons, trees, wind, fire, electromagnetic fields" (Bennett 2010, 24).

The *Forest City* metaphor offers a glimpse through the ineffable dimensions that form the core of urban intelligence. Research-creation calls upon more radical grounds to begin thinking of smartness beyond dualism, a smartness more intelligent than human thinking, machine thinking, and biological thinking alone. Much more than the sum of the parts – a regenerative principle capable of connecting all forms of thinking, known and unknown. Such a principle lies at maximum distance from abstract thought: it is rooted, aggregates, and relies upon the specific properties and vibrancy of materials, actants, places, and forms.

Image credits

All images made by the author.

References

Barua, Maan, and Anindya Sinha. 2022. "Cultivated, Feral, Wild: The Urban as an Ecological Formation." *Urban Geography* 44 (10): 2206–27. https://doi.org/10.1080/02723638.2022.2055924.

Bennett, Jane. 2010. *Vibrant Matter: A Political Ecology of Things*. Duke University Press. https://doi.org/10.2307/j.ctv111jh6w.

Cronon, William. 1996. "The Trouble with Wilderness: Or, Getting Back to the Wrong Nature." *Environmental History* 1 (1) (January): 7–28. https://doi.org/10.2307/3985059.

Guattari, Félix. (1989) 2000. *The Three Ecologies*. Translated by Ian Pindar and Paul Sutton. The Atholone Press.

Haraway, Donna. 1988. "Situated Knowledges: The Science Question in Feminism and the Privilege of Partial Perspective." *Feminist Studies* 14 (3): 575–99. https://doi.org/10.2307/3178066.

—. 1991. "A Cyborg Manifesto." In *Simians, Cyborgs and Women: The Reinvention of Nature*. Routledge.

Hörl, Erich. 2017. *General Ecology. The New Ecological Paradigm*. Edited by Erich Hörl with James Burton. Bloomsbury.

Jankowski, Patricia, Anja Höfner, Marja L. Hoffmann, Friederike Rohde, Rainer Rehak, and Johanna Graf, eds. 2023. *Shaping Digital Transformation for a Sustainable Society. Contributions from Bits & Bäume*. Technische Universität Berlin.

Kohn, Eduardo. 2013. *How Forests Think. Toward an Anthropology Beyond the Human*. University of California Press. https://doi.org/10.1525/9780520956865.

Ladyman, James, James Lambert, and Karoline Weisner. 2013. "What Is a Complex System?" *European Journal of Philosophy of Science* 3 (June): 33–67. https://doi.org/10.1007/s13194-012-0056-8.

de Lange, Michiel, and Martijn de Waal, eds. 2019. *The Hackable City: Digital Media & Collaborative City-Making in the Network Society*. Springer. https://doi.org/10.1007/978-981-13-2694-3.

Mattern, Shannon. 2017. "A City Is Not a Computer." *Places Journal*, February. https://doi.org/10.22269/170207.

Moujan, Carola. 2015. "Augmenting the Bench." *Interstices: The Urban Thing* 16 (16): 47–56. https://doi.org/10.24135/ijara.v0i0.491.

—. 2023. "The Edge of Experiences. Towards a Material Ecology of Augmented Interiors." *Interiors* 12 (2–3): 307–29. https://doi.org/10.1080/20419112.2023.2171640.

Moujan, Carola, and Agustín Ortiz Herrera. 2022. "Artistic Residencies as Critical Research: Entangled Methodologies for Future Science." *Proceedings of ISEA 2022: Possibles*, edited by Pau Alsina, Irma Vilà, Susanna Tesconi, Joan Soler-Adillon, and Enric Mor. ISEA.

Mullins, Paul D. 2017. "The Ubiquitous-Eco-City of Songdo: An Urban Systems Perspective on South Korea's Green City Approach." *Urban Planning* 2 (2): 4–12. https://doi.org/10.17645/up.v2i2.933.

Mumford, Lewis. (1934) 2010. *Technics and Civilization*. The University of Chicago Press.
Pickering, Andrew. 2009. *The Cybernetic Brain. Sketches of Another Future*. The University of Chicago Press. https://doi.org/10.7208/chicago/9780226667928.001.0001.
Rykwert, Joseph. 2002. *The Seduction of Place. The History and Future of Cities*. Vintage.
Simondon, Gilbert. (1958) 2017. *On the Mode of Existence of Technical Objects*. Translated by Cécile Malaspina. University of Minnesota Press.
Springgay, Stephanie, and Sarah E. Truman. 2018. *Walking Methodologies in a More-Than-Human World: WalkingLab*. Routledge.
Viveiros de Castro, Eduardo. (2009) 2014. *Cannibal Metaphysics. For a Post-Structural Anthropology*. Edited and translated by Peter Skafish. Univocal.

26 Unfamiliar Convenient

CLAIRE GLANOIS AND VYTAUTAS JANKAUSKAS

Artists' Statement

Caught between sheen, religion, and presence,
Through the meanders of an inaccessible brain,
The cosiness, better composed,
That show-off of a Raspberry Pi,
Pulled out of a sibilant rainbow.

A voice assistant recites a haiku inspired by the trajectories of a vacuum cleaner, offering an esoteric ritual of spatial reading. Spread through the machine's choreographies, wisps of incense dissipate in the air, carrying blessings. As the vacuum cleaner returns to its docking station, archetypical home scents – carpets indulging the presence of pets, wallpaper infused with the smell of food – slowly regain ground. Now, humans are free to walk around, restoring conventional domestic hierarchies.

The smartification of homes has been driven by the promise of personalized consumer convenience and indispensably frictionless home experience, freeing up human time and embodying the ideal that "the home is where everything else stops and you begin" (Batchelor 2017). However, *smart* domestic realities have more often been captured by anecdotes than promises of indispensability. Consider viral videos of Roomba autonomous vacuum cleaners ridden by dogs in shark suits, the melodramatic funerals given to Jibo kitchen robots, or the late-night barks of Amazon Alexa, spooking its users. Attempts to justify these purchases often escalate, retrofitting the home to accommodate the object. We rearrange cables and furniture to prevent a Roomba from getting trapped. We replace traditional light bulbs with *smart* ones, only to find ourselves raising our voices and repeating commands for Siri to switch them on. The unnatural pause, silence, before the voice synthesis responds.

Humans try to match machine limits, rather than the other way round. In contrast with the promised efficiency, our interactions with today's household devices mirror the care we provide to domestic pets. It is the convivial theatricality of the Internet of Things – relation of attachment, patience, and care – that validates their presence in our homes, much more than their marketed utility (Sterling 2014).

We propose a redefinition of domestic smartification, emphasizing continuous exploration, interdependencies, and machine-native behaviour. The *Unfamiliar Convenient* project documents the peculiar relationship between a vacuum cleaner and a voice assistant as part of a wider ecosystem of a home and its inhabitants. Sensors, processing units, and outputs such as speech synthesis are metaphorized as *species denominators* (Stiegler 1998), thanks to which devices can perceive their habitat and establish new social structures and raisons d'être.

Imagine if our everyday devices were on an open-ended quest to explore and understand our homes. This could support a more varied and nuanced set of relationships with smart devices. A voice assistant's ability to verbalize would not be limited to a mimicry of human speech. Vacuuming, reframed as an excuse to roam and wander, could move beyond topographic optimization, facilitating surprising encounters with humans, pets, or objects.

A range of concepts could bring new *smartnesses* into practical use. *Sensitivity*, or a vacuum cleaner's response to stimuli, such as collisions, could influence the device's carefulness, fostering behaviours conducive to interspecies play. *Adaptability*, tracking the appropriation of such devices by individuals and communities (Kudina 2021), could help devices settle where top-down approaches struggle, such as acknowledging cultural specifics in the use of technology like minor dialects or religious customs. *Historicity* (Longo 2015) could support the archival storage and exchange of context-specific information among devices within specific settings, as opposed to standardized cloud services. A more situated, localized approach could, in turn, affect devices' *resilience*, with features (energy consumption, networking, etc.) adapting to the pressures of climate and infrastructural volatility.

This project contributes to the field of human-computer interaction by *rethinking smartness* through more open-ended approaches (Porfiri et al. 2021). Imagine these devices placed in volunteer households for extended observation, documenting their operations and use, and interviewing their keepers. This would provide insights into why the promises of *smart* technology, which aim to streamline our domestic lives – whatever that would mean – so rarely become material realities (Hester and Srnicek 2023).

Rather than striving for seamless integration and control, the project advocates for coexistence, where machines are allowed to exhibit their unique behaviours and adapt to the ever-changing dynamics of domestic life. If machines are free to participate in our home ecologies – where we already negotiate intimacy and embrace volatile rule sets – we humans can be more open-minded in our interactions. Instead of focusing solely on a given machine's utility or potential to replace human labour, we could approach it with curiosity, asking ourselves what we might learn or discover.

The behaviours of these two experimental (yet functional) objects have been documented in a short film, available via https://vimeo.com/641369896. By presenting the vacuum cleaner and voice assistant as subjects of observation, akin to wildlife, the film challenges the conventional notions of domesticity, encouraging a more exploratory relationship with technology.

References

Batchelor, L. 2017. "A Better Life for the Many People." *REAL Review*, no. 5, 14–29. https://real-review.org/home.

Hester, H., and N. Srnicek. 2023. *After Work: A History of the Home and the Fight of Free Time.* Verso Books.

Kudina, Olya. 2021. "'Alexa, Who Am I?': Voice Assistants and Hermeneutic Lemniscate as the Technologically Mediated Sense-Making." *Human Studies* 44 (June): 233–53. https://doi.org/10.1007/s10746-021-09572-9.

Longo, Giuseppe, Maël Montevil, Carlos Sonnenschein, and Ana M. Soto. 2015. "In Search of Principles for a Theory of Organisms." *Journal of Biosciences* 40 (5): 955–68. https://doi.org/10.1007/s12038-015-9574-9.

Porfiri, M., R. Halverson, F.J. Vazquez-Abad, N. Kozlova, J. Strang, S. Moturu, et al. 2021. "Assessing Open-Ended Human-Computer Collaboration Systems: Applying a Hallmarks Approach." *Frontiers in Artificial Intelligence and Applications* 4 (October): 670009. https://doi.org/10.3389/frai.2021.670009.

Sterling, Bruce. 2014. "The Wolf Is in the Living Room: Excerpts from a Conversation with Bruce Sterling." In *SQM, The Quantified Home*, by Space Caviar. Lars Müller.

Stiegler, Bernard. 1998. *Technics and Time 1: The Fault of Epimetheus.* Stanford University Press. https://doi.org/10.1515/9781503616738.

27 Conclusion

KELLY BRONSON, VINCENT MIRZA, AND
MASCHA GUGGANIG

We know now that smart places and objects are often obscuring more complex realities articulated by complex social relations, fissures, and frictions, all of which exceed the smartification envisioned by urban planners, engineers, and developers. The contributions in this book have shown us that hegemonic notions of smartification qua urban digitization do not necessarily produce the sleek and seamlessly integrated whole that is promised. And yet, in many cases a kind of modern smart city nonetheless results from a performance of a smartness rhetoric that helps to produce a specific technological world and to obscure alternatives to this world. The contributions in this book have brought a critical perspective on the "smartification of everything" and offered the reader alternative enactments of smartification – more nuanced and complex portraits of contemporary relations, whether city-country, human-machine, human-nature, or Global North-South. The contributions have also revealed the discursive utility of hegemonic notions of smartness – such as smart city competitions as a means to control uncertain postcolonial futures. Finally, they problematize the dominant smartification rhetoric for its impact on places and people at the margins.

We saw how in both Jordan and Myanmar, smartification layers onto historic patterns or entrenched inequities and extraction. In these places, smartification is brought into being within landscapes long marked by colonial and imperial dispossession – places where historic dispossession, like contemporary smart infrastructures, concentrates wealth and exacerbates landlessness. Myanmar telecom towers that enable smart farming apps, for example, were built in rural landscapes characterized by decades of military land-grabbing. Many of the book's essays focus on smartification in the Global South – from Myanmar to Pakistan – and reveal that smartification positioned as global "development" is contentious, incomplete, and contested by locals. As smart technologies are adopted and implemented, they encounter the frictions of lived realities, such as those of people on the sidewalk in Pakistan, who are devising alternative processes for temperature control and living with the realities of climate change. Said plainly, there is the ideal (often formulated in a western, technocratic imagination) of smart cities, and then there is the reality of smartification as a complex social and cultural process, not simply a technical one.

Beyond just expanding the site for research on smartification to the Global South, the essays in this book have expanded analyses of smartification beyond the city, focusing instead on interrelationships: between city and country, between plants and technology, between people and domestic robots. Some essays focus on smart city projects in prominent cities such as Vienna, but even these reveal how smartification processes are non-linear and always encompass a reconfiguration of relationships among non-human actants (surveillance cameras, etc.) and human inhabitants. Many of the essays also reveal the hidden but necessary substrate supporting smartification – the mineral resources required for digital infrastructure (Fard), or the social capital among municipal wastewater workers in downtown Houghton, Michigan (Hardy). The essays and especially the artist contributions also showcase the crucial role of mundane objects – from scooters to traffic lights.

The book has also revealed the importance of inter- and transdisciplinary dialogue, specifically the articulation of art with more traditional academic writing for fostering critique and critical conversation. When we started planning for the symposium that began this book project, which was also called "The Smartification of Everything," we wanted to establish a dialogue between various academic disciplines and a platform for the visual and mixed-media arts through an exhibition entitled *smART*. This curated show was a testing ground for various conceptions of technoscientific smartness and smartification processes, and it was a playground that engaged members of the wider public in this conversation. In *Critical Zones*, Latour and Weibel say that the arts or "exhibition" scales the problems we are trying to think about, such that "today, much as in other earth-shaking periods, we need aesthetics" (2020, 8) to render us sensitive to alternatives and other ways of being. Raymond Williams (1977) suggests that the arts allow us to see what is emergent in society and to provide a view into the near future. This book recreates this interdisciplinary and arts-led conversation for the reader. In doing so, we offer a pedagogical approach that might be recreated by others. *smART* has been a critical tool for revealing the contradictions and frictions inherent in smartification processes when applied across various settings – from rural environments to cities in the Global South.

The interdisciplinarity and the arts-academia collaboration modelled in this book builds new networks and worlds for thinking and doing differently in our attempt to live well together in humans-machines-nature in the twenty-first century. Just as smartification is always incomplete, we invite you to continuously rethink smartification practices through a regular process of critique, reflection, and engagement with diverse stakeholders.

References

Latour, Bruno, and Peter Weibel, eds. 2020. *Critical Zones: The Science and Politics of Landing on Earth*. MIT Press. https://doi.org/10.1162/leon_r_02159.

Williams, Raymond. 1977. *Marxism and Literature*. Oxford University Press.

Contributors

Martin Abbott is a lecturer in urbanism at the University of Sydney. His research examines how technoscience shapes human-environment relations, particularly in urban contexts characterized by uncertainty and inequity. In his current research project, titled "Hybrid Landscapes," he follows public officials and experts as they navigate New Orleans's future on the deltaic floodplain of the Mississippi River and the ongoing transformation of flood risk-related knowledge in the United States. Martin holds a PhD in science and technology studies from Cornell University.

Sebastian Bornschlegl is a PhD candidate at the Chair of Sociology of Technology, Risk and Environment at Stuttgart University. His research centers on urban assemblages of human and non-human forces under development pressures. His master's thesis on the reassembly of mobility for automated public transport takes place in Seestadt (Vienna), just like the artwork he presented at the smART exhibition.

Kelly Bronson is the Canada Research Chair (II) in Science and Society at the School of Sociological and Anthropological Studies and the Institute for Science, Society & Public Policy at the University of Ottawa. Her research explores the social and ethical dimensions of emergent technology like big data and AI. Besides her forthcoming book *The Immaculate Conception of Data: Agribusiness, Activists, and Their Shared Politics of the Future* (McGill-Queen's University Press), her work has been published in journals such as *Big Data and Society*, *Science as Culture*, and *Journal of Responsible Innovation*.

Hannah Carlan is a cultural and linguistic anthropologist who is currently an applied researcher studying patient experience in the US healthcare industry. Since earning her PhD in anthropology from UCLA in 2021, she served as a visiting assistant professor of anthropology at Rollins College, an Experience Research Director at Elevance Health, and is currently Senior Manager of Insights at Collective Health. Her academic research has examined statemaking, welfare, and rural development in Himalayan India, with particular emphasis on the constitution and contestation of rural women's political agency through everyday linguistic and semiotic practices. She has published in the *Journal of Linguistic Anthropology, International Journal of the Sociology of Language, Anthropology News, Texas Linguistic Forum*, along with multiple edited volumes.

Laure Dobigny is an associate professor with ETH+ – ETHICS EA 7446, Université Catholique de Lille, F-59000 Lille, France, where she works on the imaginaries of energy transition. She is also an associate researcher at the Centre for Administrative, Political and Social Studies and Research (University of Lille). She holds a PhD in socio-anthropology (University Paris 1 Panthéon-Sorbonne).

Gabriel Dorthe is a senior scientific assistant at the Swiss Federal Institute of Technology (ETH) Zürich, Department of Humanities, Social and Political Sciences. He works on public mistrust in science and science activism. He is also an associate researcher with ETH+ – ETHICS EA 7446, Université Catholique de Lille, F-59000 Lille, France. He holds a PhD in philosophy (University Paris I Panthéon-Sorbonne) and environmental humanities (University of Lausanne).

Bobbie Fan is a data monkey ruining the banquets of their former homes in academia and local government. They channel institutional knowledge and complicity into grassroots work against techno-solutionism, settlement, and carcerality. They are a community researcher with Pittsburghers for Public Transit and the Abolitionist Law Center, and they organize with Against Carceral Tech (ACT – formerly the Coalition Against Predictive Policing Pittsburgh & CMU Against ICE) and the coveillance counter-surveillance collective.

Ali Fard is a designer, researcher, and educator, currently assistant professor of architecture at the University of Virginia. His research and creative work explore the uneven spatial geographies of technology and infrastructure.

Aqdas Fatima is a Karachi-based designer and researcher interested in the social, political, and environmental dimensions of urban transformation. She holds a master's degree in design & urban ecologies from

Parsons, The New School, and a bachelor's degree in social and biological anthropology from University of Kent, UK. Her current work investigates how grassroots repair practices in NYC challenge dominant narratives of urban innovation and offer alternative frameworks for sustainability and care in rapidly transforming cities.

Hilary Faxon is a human geographer and political ecologist working on environment, development, and technology with a focus on social justice in the Global South. She is an assistant professor of environmental social science at the University of Montana.

Claire Glanois has a PhD in mathematics and is working as a postdoctoral researcher at the IT University, Copenhagen. Her current research is concerned with artificial intelligence, from automated decision-making to artificial life, and open-ended evolution.

Mascha Gugganig is a socio-cultural anthropologist, an STS scholar, and a curator for research exhibitions on contested (agri-food) technologies, and science-society relations. She is a lecturer and co-director of the Center for Life Sciences and Society at the Ludwig Maximilians University of Munich (LMU), and will soon start a position as senior lecturer at the Department of Science, Technology and Society at the Technical University Munich. Her work is published in *Science, Technology, & Human Values, American Anthropologist*, and *Agriculture and Human Values*, among others.

Jean Hardy is an assistant professor of media & information at Michigan State University, where he directs the Rural Computing Research Consortium. His research focuses on the role of technology in rural economic development, as well as the use of social technologies for community building among LGBTQ+ people living in rural areas.

Vytautas Jankauskas is an artist and designer intrigued by the visual and sociocultural dissonances brought about or amplified through consumer technologies. He currently serves as the director of Digital Pool at HEAD – Genève (Geneva University of Art and Design).

June Yeoreum Kim is a media and technology scholar trained in the Department of Media, Culture, and Communication at New York University. In her work, June critically engages with how technologies make humans feel, behave, and make relations, and how this human-technology interaction co-constitutes changing forms of global techno-capitalism in the digital era. June's inquiries are grounded in her positionality as a US-trained, Korean-born person, observing contemporary Vietnamese society.

Kendra Kintzi is a human geographer and doctoral candidate at Cornell University, where her research focuses on uneven development and environmental change in the Middle East. She draws together theory and methods from political ecology, postcolonial feminist geography, and digital geography to question the relationship between energy and political power, asking how urban communities in Jordan experience and shape digital infrastructure and environmental change.

Soha Macktoom is an associate director at the Karachi Urban Lab at IBA University, Karachi, and lecturer at the Social Sciences and Liberal Arts Department (SSLA) at IBA. She is an architect, with a master's in urban and regional planning from NED University. Her work looks at the physical transformations of the built environment and how architecture, design, and history are tools for understanding the city's informal settlements, urbanization trends, infrastructure, and climate. Her work has been published in academic journals including *Urban Studies, City, Future Anterior,* and *Economic and Political Weekly* (EPW).

Sarah Marquis is an activist and scholar living on unceded Algonquin territory in Ottawa, Ontario, Canada. She recently completed her PhD in environmental sustainability at the University of Ottawa. Her academic research focuses on the role of digital technologies, like robotics and AI, in transitions to environmental sustainable agriculture in Canada. Broadly, she applies a critical lens to technologies that are framed as solutions to environmental problems in agriculture. She now works for the National Farmers Union in Canada, advocating for just and sustainable agricultural policy that supports farmers and farm workers.

Vincent Mirza is the uOttawa-ULyon Joint Research Chair on Urban Anthropocene at the School of Sociological and Anthropological Studies at the University of Ottawa. As an urban anthropologist, his research is concerned with the relationship between cities and the environment in order to conceptualize the issues and challenges of an anthropology of contemporary worlds in Japan, East Asia, and in large urban agglomerations.

Bruno Moreschi is a researcher and multidisciplinary artist. He is a postdoctoral fellow at the Faculty of Architecture and Urbanism at the

University of São Paulo (FAUUSP), PhD in Arts at the State University of Campinas (Unicamp), with a Capes scholarship, and exchange at the University of Arts of Helsinki (Kuva Art Academy), Finland, via CIMO Fellowship. His investigations are related to the deconstruction of systems and the decoding of social practices in the fields of arts, museums, visual culture, and technologies. His projects have been recognized by ZKM, Van Abbemuseum, thirty-third Bienal de São Paulo, Rumos Award, Funarte, Fapesp, University of Cambridge, and CAD+SR.

Carola Moujan is an artist, designer, researcher, and assistant professor at GOBELINS, Paris (France). Her practice spans across interaction design, digital art, installation, artist's editions, urban design, scenography, and cartography. Her research through design work explores the entanglement of interactive processes connecting spaces, temporalities, sensitive bodies, affects, materials, and digital code in more than human contexts.

Junnan Mu is a PhD candidate in African and African American studies and a fellow at the Film Study Center at Harvard University. She combines an ethnographic and multisensory approach to understanding urbanism, digitality, spirituality, and the technoscientific future-making in Kenya and beyond. Her work is supported by the Wenner-Gren Foundation of Anthropology.

Gabriel Pereira is an assistant professor in AI and digital culture at the University of Amsterdam (UvA), based at the Media Studies Department and the Institute for Logic, Language and Computation (ILLC). His research focuses on critical studies of data, algorithms, and digital infrastructures, particularly those of computer vision. Projects with Gabriel have been exhibited in venues such as the thirty-third São Paulo Art Biennial, the Van Abbemuseum, IDFA DocLab, and Itaú Cultural. He is a researcher in residence at the Center for Arts, Design, and Social Research (CAD+SR). www.gabriel pereira.net/.

Simon Rabyniuk is an architectural researcher and designer based in Toronto. His research and teaching develop historical and theoretical perspectives about technology and urbanization. He is currently a PhD candidate at TU Eindhoven in architectural history and theory, and he serves as a sessional lecturer at Daniels Faculty of Architecture, Landscape, and Design. From 2010 to 2015, he was principal at the research, art, and design studio Department of Unusual Certainties. He holds a professional graduate degree in architecture from the University of Toronto (2019), where he was the recipient of multiple awards.

Pouya Sepehr is an academic researcher with a background in science and technology studies, urban sociology, and innovation studies. His academic journey has been largely anchored at the University of Vienna, where he has contributed as a university assistant and lecturer within the Department of Science and Technology Studies. His research portfolio is broad and multifaceted, spanning areas such as smart cities, urban programming, critical algorithm studies, spatialized technologies, smart communities, urban development practices, and urban imaginaries. Beyond his academic pursuits in Vienna, Pouya has been a visiting fellow at the European Ethnology Department at Humboldt University of Berlin. Currently, he is engaged in research at the Institute for Advanced Studies (IHS), Vienna. In addition to his research and teaching roles, Pouya is an active council member of the Society for Social Study of Science (4S), further demonstrating his commitment to the academic community.

Devin Shepherd is a writer; he graduated from the University of Arkansas with a master's degree in English.

Abhishek Viswanathan made his way to Pittsburgh from a middle-class, privileged-caste house in Mumbai, India, a burgeoning metropolis, to pursue graduate studies. With an undergraduate degree in engineering, and a graduate degree in Telecommunications, a big part of his educational trajectory has been within the STEM realm. However, being an immigrant participating in labor organizing, environmental activism, and anti-colonial solidarity work has led him to explore perspectives from the social sciences, specifically critical discourses on techno-solutionism, ongoing settler-colonial and racial capitalist frameworks that color the technology development landscape today. As an assistant professor at Chatham University, he continues to grapple with the many contradictions within academia, and the many intersections and impacts that must be considered in our research.

Juliette Walker is an interdisciplinary artist; she graduated from the University of Arkansas with an master's degree in Fine Arts in Studio Art.

S. Ashleigh Weeden, MPA, PhD, is an award-winning rural futurist and community-focused researcher. Her work focuses on the way people, place, and power dynamics are reflected in and impacted by policymaking. Recognized as a thought leader in the growing body of research and advocacy on "the right to be rural," Ashleigh works to advance evidence-based policy and public sector leadership across a wide variety of critical portfolios, including infrastructure, innovation, and inclusive community economic development. Her work can be read in publications like *The Right to be Rural*, the Canadian Rural Revitalization Foundation's *State of Rural Canada and Rural Insight Series*, *The Conversation Canada*, *CIGI Online*, *Policy Options*, and *Municipal World*.

Meg Wiessner is a researcher and artist based in New York. She investigates relationships between computational media, design, and architectural materials. Her doctoral research asks how digital technologies shape the emerging political ecology of mass timber in the Pacific Northwest.

Index

abolitionist practice: care, 53; collectivity, 53; refusal, 53
actors: digitization and smartness, 3; local, 21–2, 91, 122–6, 269–70; multilateral, 33, 41; national and private-level, 38, 67–8; non-human, 138, 151; smartification and, 108–9, 133–4, 166, 222; state/non-state, 111–12, 245–9, 252; tech disruptors as development actors, 37–8; technoscientific, 5–6. *See also* users
adaptability, xiii, 286
affect, 13, 58, 152, 269, 276
agriculture: agroecology, 122, 126–7; food sovereignty, 122; optimized, 120–6; precision, 120–5 (*see also* smart farming); smart, 11, 16, 33, 121–4; smart farm, 121, 126; sustainable, 122–5; sustainable intensification, 122
AI (artificial intelligence), 4, 8, 17, 99, 119, 124, 139, 192, 208
algorithms: capabilities, 264; epistemology of, 8–9, 151, 208–11; equity and, 199–203; governance and, 34–6; inequality and, 191–2, 271–5; political, 21, 51; technology of, 72, 271–5, 99, 119–21, 203, 243. *See also* prediction

animal graveyard, 162–4
anthropocentrism, 9–10, 18
apps, 37–9, 100, 289
architecture: low-carbon design, 53–7; mass timber, 57–66; optimized, 120–6; smartification of, 15, 95–9, 137–9; solid nature of, 109; vertiplaces, 16, 107–16
art, definition of, 12, 244
art, science, and technology studies (ASTS), 5, 12
artist-researchers and artist-scholars, 5–7, 11–15, 18–19, 21, 133–4, 243, 272
asset management, 17, 193, 231–9
austerity and governance, 33–7

bias, 5, 8–11, 17, 21, 191–2
black box (of technoscience), 14, 17, 97
boundaries: geographic and space, 14, 17, 111, 139–40, 191; infrastructures, 151–2, 155–66; political, 114; smart actors, 133–6, 255
broken windows theory, 49–50
building: city, 10, 34, 81–2; expansion, 108–11, 114, 141–4, 231–2, 238, 245–9, 254–7; geographical, 18,

68–75; modelling, 57; political, 31, 156, 160, 243, 245–9; smartness and, 3–4; smart users in, 95–104; university, 16

capitalism: colonial/racial, 13–14, 32, 36, 45–6, 126–7; digital, 81–8, 225, 264–6; global, 76; platform, 139–40
cars: country, 113–6; as inventions, 255; pedestrian crosswalks and, 209–12; self-driving, 8–10, 48–52; smart cameras and, 17–18, 243, 264
cities: city-region, 108–16; creative, 14, 32, 45, 53; dumb, 20; forest, 18, 244, 271–81; gentrification in, 48–53; smart, 5–10, 15–22, 33–49, 68–77, 81–97, 112–16, 134–6, 138–47, 155–66, 208–16, 244–59; synonyms for, 83–5, 88–91 (digital, cyber); test-bed urbanism in, 68, 71–4, 140; urban fantasies of, 86–93; urban laboratories in, 20, 45, 51. *See also* smart urbanism
colonialism (settler), 13–14, 32–3, 37, 46, 66, 122
computational urbanism, 207
connectivity, 34–40, 97–8, 116, 207
cooling technology: thermal properties and injustice and, 248–59; vernacular vs. smart solution and, 15, 18–20, 98, 243–57
cosmopolitanism, 88–9, 91
creative class, 14, 32, 45
critical development studies (CDS), 35–6, 40
cyborgs, 21

definitional impasse, xiv, 15, 83–91, 256
demo-logics of smartification, 5, 10
demystification, 52
deterritorialization, 272–80

disability and accessibility, 209–16
disposability, 155–66
drones, 16, 108–14

eco-logical forms of thinking, 19, 270–2
economic growth and smartification, 9, 33–6
education and universities as engines of smart venture, 15, 32, 45
epistemology of smartness, 5–6, 35, 151–2, 191
erased land, 15, 32, 68–77
ethics of smartness: moral dimensions/considerations, 7–9, 15–19, 208, 213; social justice, 3–4
extraction: data, 76, 121, 145–7, 265; land and natural resources, 10–16, 21, 31–41, 45–51, 57–60, 76, 126, 134–40

farming. *See* agriculture
fissures, 81–2, 84–5, 121, 289
fixity, spatial, 139
forestry, climate-smart, 31–2, 58
frictions, 14–15, 81–2, 85, 108, 155–6, 289

gentrification, 48–53
governance: digital, 155–66; national and global, 122; smart, 34–7, 207–15; urban, 137–8, 208, 221–7, 247–52
governmentality, 101, 156

hybrids, 13, 58

imaginaries: collective, 96; infrastructures, 20, 32–4, 40, 48–50, 86–7; policymaking, 125; sociotechnical, 5–6, 10; urban centered, 90, 191–7, 247
Internet of Things (IoT), 6–8, 33–4, 244, 273–8, 286

knowledge-based economy, 45–6

land clearing, 15, 32, 68–77
land transmogrifying, 47
legibility, 111
liminality, 84
living thought, 277
luxury, 76–8

materiality of smartification, 5, 15, 39, 71, 171
mobility, framing, 115
monofunctional vs. ecological perspectives, 112–15

narcissism, geographic, 220–7
necropolitics of disposability, 155–63
neoliberalism, 15, 17, 21, 32, 46, 85, 225
normativity and heterogeneity. *See* frictions

optimization: of everything, 5–6, 19, 68, 286; smartness, 120–6; sustainable, 16, 138; technoscientific, 9

postcolonial smartification, 33–7
power relations: and authority, 46, 122, 213–27; and dynamics, 76–8, 81, 156–60; of communities, 51, 68–72; inequality in, 252–72, 281, 289; and technology, 3–17, 33–41, 140–7
prediction. *See* algorithms
platform capitalism, 140

race, racial capitalism and smart systems, 15, 32–6, 45–6
registers, ecological, 270
rethinking smartness, 243, 258, 286–7
revitalization, 47

rubble, 76–8
rurality, 192, 219–27

science and technology studies (STS), 5, 7, 12–17, 96–9, 133–4, 196
scooters, 17, 134–5, 169–70, 290
semiotics: infrastructure, 58, 166; processes, 90–2, 164, 271–9; of smartness, 9, 84 155–6
simplification, 82, 109–15
smart: as a buzzword, 15, 32, 35; defined, 3, 15, 20, 243; epistemology, 4–5; meanings of, 7, 9. *See also* agriculture; cities; climate; cooling technology; education; race; smart people; smartness; urbanism; users; waste
smART, 7, 11–13, 290
smart cities. *See* cities
smart dustbin. *See* animal graveyard
smart farming, 11, 16, 21, 39, 100, 119–26, 289. *See also* agriculture, precision
smart people, 88–91, 98
smart technology: infrastructures, 70, 123, 233, 286; and smart users, 82, 96; STS, 196–9, 208–13; sustainable, 123
smart urbanism, 137–47, 170, 208, 212, 216, 269. *See also* cities
smartification: academic, 32, 45; agendas, 21, 192–3, 219–25; definition of everything, 6–8, 207, 214, 289–90; discourse of everything, 220; materiality of, 5, 15, 39, 71, 171; postcolonial, 33–7; practice and process of, 10, 16–17, 21, 31–4, 41, 120, 133; roots of, 31. *See also* actors; architecture; demo-logics; economic growth; materiality; postcolonial; technology; wal*smartification

smartmentality, 222–7
smartness: academic production of, 45–6; actors and technoscientific, 133; biological, 244, 270; as empty signifier, 7, 15–16, 88, 91; epistemology of, 5–6, 35, 151–2, 191; ethics of (*see* ethics of smartness); definitional impasse, 12–13, 18–22; idiotic (unsmart) as contrast, 18–20, 246–8, 254–9; mandate, 119–21; mediated space in, 77–8, 137, 211; rearticulation/rethinking of, 18–19, 20–1, 243, 258, 286–7; technoscientific, 4, 6–8, 32, 81, 290; urban, 18–19, 244, 269–80; visionaries of, 5. *See also* academic production; actors; biological; epistemology; ethics; rethinking; sustainable; visionaries
sustainable smartness, 25. *See also* agriculture; sustainable

techno-solutionism, 46–7, 88–91, 209–15
technology: cohabitants, 19; fetishization of, 8–14, 58, 91; transformative role of, 35–40, 84, 279–80; technocentred practices of smartification, 270. *See also* smart technology
test-bed urbanism. *See* cities
traffic lights, 18, 192, 208–16

unsmart. *See* smartness
urban: acupuncture, 89; enablers, 208; laboratories (*see* city); liveability, 157–66, 247, 258 (*see also* cities); usability, 157–66, 166 (*see also* cities); mobility, 107–16
users of smart systems, 15, 82, 95–104. *See also* cities

vertiplaces, 16, 107–16
visionaries of smartness, 5
visualization and aesthetics, 5–6, 18–20, 236–7, 245–57, 263–6, 290–1

wal*smartification, 134–5, 170–87
waste and smart infrastructure, 10–13, 122–3, 155–66, 192–3, 231–7